AF556676

# Ensuring Sustainability in Forestry

# Ensuring Sustainability in Forestry

## Certification of Forests

H S Gupta • M Yadav • D K Sharma • A M Singh

The Energy and Resources Institute

ISBN 978-81-7993-495-1



**Suggested citation**
Gupta, H. S., Yadav, M., Sharma, D. K., and Singh, A. M. 2013. *Ensuring Sustainability in Forestry: certification of forests*. New Delhi: TERI

**Published by**
The Energy and Resources Institute (TERI)
TERI Press
Darbari Seth Block
IHC Complex, Lodhi Road
New Delhi – 110 003
India

***Tel.*** 2468 2100 or 4150 4900
***Fax*** 2468 2144 or 2468 2145
India +91 • Delhi (0) 11
***Email*** teripress@teri.res.in
***Website*** www.teriin.org

Printed in India

# Foreword

In view of the current international concerns, threats to natural ecosystems and biological diversity, deforestation, forest degradation, and multifaceted effects of climate change, certification can prove to be an important tool to ensure sustainable use of forest resources. It comprises many elements pointing to a new form of multi-stakeholder governance, which is also expected to lead to better governance and satisfaction of the user's needs.

Forest certification is an innovative approach to independently verify the quality of forest management and to communicate this to market players to improve market benefits for the products developed through good management. Certification provides a guarantee to buyers that their purchase comes from well-managed forests that do not favour unsustainable or inequitable practices, which are the main cause of forest problems. But more important is the role of certification in providing an incentive to forest owners and managers to raise the standard of their forest management. Current certification initiatives are developing against a rapidly changing background of international and national initiatives in forestry, biodiversity, conservation, environmental management systems, climate change considerations, and trade—many of which also aim to achieve sustainable forestry on the ground.

There are increased environmental measures and demands for forest certification in global markets to ensure that trade in wood and wood-based products is sustainable in accordance with their national, regional, and multilateral environmental policies. The book *Ensuring Sustainability in Forestry: certification of forests* aims to discuss forest certification in relation to the associated opportunities and constraints. Besides the basic concepts, processes, spread, challenges, and other issues like convergence, impact on biodiversity, and implications on the Kyoto protocol, the book explains clearly how a certification programme should function in the Indian context. It discusses critically what certification may or may not achieve in terms of solving the problems of forests. In addition, it gives an up-to-date account of the relationship between certification and other efforts being made to improve forest management worldwide.

This book is practice oriented in nature. I congratulate the editorial team and contributing authors for their excellent work. We hope that this book will serve as a useful guide to professionals serving in government, private, and non-governmental organizations, helping them to fully realize the potential of sustainable forestry projects, which will consequently deliver certified products and services along with achieving sustainable development, particularly in the Indian context. Since the majority of Indian forests are under the government sector, which are already subject to fairly strong checks and balances from different wings of the state apparatus and the working system is built on detailed plans based on sustained yield, the general principles and criteria will have to be modified considerably to take into account these realities. At the same time, the forest departments would need to undertake a systematic procedure of subjecting their processes and practices to detailed scrutiny to make them more efficient, transparent, and user friendly. A process of capacity building jointly with users, local communities, and other stakeholders will be needed to achieve this.

**P J Dilipkumar**
Director General (Forest)
Ministry of Environment and Forests (MoEF)
Government of India
New Delhi

# Preface

Forest managers all over the world have been under increasing pressure to meet the growing demand for all types of forest products and services, while striving to maintain the ecological integrity of forest ecosystems and enhance the livelihood of people on a sustainable basis. Many countries of the world, including those in the Asia–Pacific region, have established a strong tradition of sustainable forest management (SFM) through their forestry and environmental wings.

In this background, the concept of sustainability in forestry is moving from mere statements of principles to the development of official definitions and practical instruments to achieve it. The standard guidelines and various criteria and indicators are being developed at various levels for forest management, auditing, and certification. These initiatives are part of a global concern for forestry raised in various international forums. More important is the role of certification in providing an incentive to forest owners and managers to raise the standard of their forest management. It is, of course, not the only force in this direction.

Current certification initiatives are developing against a rapidly changing background of international and national initiatives in forestry, biodiversity, conservation, environmental management systems, and trade—many of which also aim to achieve sustainable forestry on the ground. Within this context, there is scope to focus on certification so that its potential contribution can be fully realized in the Indian context. In order to meet the growing demand for forest certification in the global market, Indian industries have to look for certified forests/plantations to source their raw materials. Forest certification as a concept is at a very nascent stage in India. The book *Ensuring Sustainability in Forestry: certification of forests* has been designed in such a way that it can facilitate the wide range of stakeholders to meet their information needs and improve their conceptual understanding within a broad framework about forest certification, which has national and societal relevance.

An effort has been made in this book to discuss various emerging issues related to forest certification in a comprehensive manner and in a lucid language. The book is expected to be a handy guide to professional foresters, planners, plantation companies, forest growers and scientists, and all those having interest in the subject.

# Acknowledgements

The authors are grateful to the National Afforestation and Eco-development Board (NAEB) under the Ministry of Environment and Forests (MoEF) for the help in the preparation and publication of this book. The authors sincerely acknowledge the guidance, support, and patronage of P J Dilipkumar, Director General of Forest, MoEF; Shri K B C Thampi, former Inspector General, NAEB; and Shri A K Singh, Principal Chief Conservator of Forests, Jharkhand Forest Department at different stages.

This work would not have been possible without the financial support provided by the NAEB, and the authors sincerely thank it for this support and for having reposed faith in the academic team of the Indian Institute of Forest Management (IIFM), Bhopal.

We also gratefully acknowledge the ideas of various academicians and practising foresters who have helped us in structuring the book and who contributed suggestions and data, which laid the foundation for the book. The authors would like to thank them for their support, encouragement, and inputs in the form of critical comments for the improvement of the book.

The authors thank Dr R B Lal, Director IIFM, for reviewing the work and offering helpful comments and suggestions and, above all, for his constant support and guidance on the overall work.

The authors are also thankful to Ms Lolita Pradhan and Ms Kalpana Basera, researchers at IIFM, who helped the authors in structuring the draft and further refining it with their untiring and sincere efforts.

We are highly thankful to TERI Press for publishing this book and, in particular, to Ms Yoofisaca Syngkon Nongpluh and Mr Arun Kumar Paul for their untiring effort in getting the work reviewed and edited before its final printing.

Thanks are also due to numerous friends and well-wishers for their formal and informal comments on this draft and on related documents. They deserve special thanks for their support in the development of this analysis and their incisive criticism at the draft level.

Our thanks go out to the various state forest departments, particularly of Jharkhand and Madhya Pradesh, and their officers who have always supported and encouraged this work.

# Contents

*Foreword* *v*
*Preface* *vii*
*Acknowledgements* *ix*
*Abbreviations* *xvii*

**1. Sustainability and Forestry** **1**
- 1.1 Concept 1
- 1.2 Emergence of the Concept of Sustainability in Forest Management 2
- 1.3 Landmark Events in Sustainable Forestry 4
- 1.4 Sustainability: concept in forestry 7
- 1.5 Sustainable Forestry: various facets 8
- 1.6 Goal of Sustainable Forest Management 11
- 1.7 Genesis of Sustainable Forest Management 12
- 1.8 Basic Requirements for Achieving Sustainable Forest Management 17
- 1.9 Efforts to Ensure Sustainability in Indian Forests 20
- 1.10 Modern Efforts towards Sustainability in Indian Forestry 21
- 1.11 Initiatives to Integrate Elements of SFM in Current Indian Forestry 22

**2. Concept and Evolution of Forest Certification** **23**
- 2.1 Meaning of Forest Certification 23
- 2.2 Need for Certification 27
- 2.3 Objectives of Certification 28
- 2.4 Emergence of Forest Certification through History 30
- 2.5 Emergence of Certification in America and Europe 31
- 2.6 Certification: an initiative to combat illegal logging 32
- 2.7 Extent of Forest Certification: status review 33

**3. Schemes for Certification** **41**

3.1 Types of Certification Schemes 41

3.2 Various Schemes for Forest Certification 45

3.3 Some National Certification Programmes 55

3.4 Comparison of Forest Certification Schemes 59

3.5 Different Standards of Forest Certification: strengths and weaknesses 59

3.6 Comparing Two Major International Schemes 63

3.7 Choosing a Certification Scheme 63

3.8 Conclusions 63

**4. Potential and Status of Certified Forest Products** **65**

4.1 Market for Certified Forest Products 65

4.2 Accessing New Markets 66

4.3 Demand for Certified Forest Products 66

4.4 Market Demand for Certified Forest Products 67

4.5 Developments in CoC Standards and Procedures 68

4.6 European Markets for Certified Forest Products 70

4.7 Status of Forest Certification in India 74

4.8 FSC Certification in India 77

**5. Process of Certification of Forests** **81**

5.1 Process of Certification 81

5.2 Elements of Forest Certification 83

5.3 Comparison of CoC Certification Schemes 93

5.4 Pitfalls and Solutions to an Effective Certification Process 96

**6. Expectations of Stakeholders** **99**

6.1 Stakeholders and their Expectations 99

6.2 Competing Stakeholder Interests 101

6.3 Customer Expectations 105

6.4 Expectations of Investor, Donor, and Insurer 106

**7. Economics of Forest Certification** **109**

7.1 Broad Economics of Forest Certification 109

7.2 Costs of Chain of Custody 111

7.3 Analysis of Cost 112
7.4 Consequences of High Cost 113
7.5 Overall Economics of Forest Certification 117

**8. Addressing Sustainability 119**
8.1 Objectives of Forest Certification and Status of Current Achievement 119
8.2 Some Obvious Effects of Certification 121
8.3 Measuring the Impact of Certification 122
8.4 Issues Related to Certification Standards 122
8.5 Forest Certification: does it really curb deforestation? 124
8.6 Does Forest Certification Really Improve Forest Management? 124
8.7 Does Forest Certification Reduce the Environmental Impact of Forestry? 125

**9. Certification of Forests and Issues of Environmental Conservation 129**
9.1 Objectives of Certification Revisited 129
9.2 Certification: a tool for conservation 130
9.3 Threats Posed to Biodiversity from Inappropriate Forestry Practices 132
9.4 Conservation Concerns in Certification Systems 132
9.5 Measuring the Impact of Certification on Biodiversity 132
9.6 Limitations in Assessing the Impact of Certification on Biodiversity Conservation: some solutions 135
9.7 Impact of Certification on Plantation Management 139
9.8 Conclusions 140

**10. Different Certification Systems: convergence issues 143**
10.1 Issue of Harmonization and Point of Divergence 143
10.2 Conflicts and/or Incompatibility between Legal Settings and Certification Standards 146
10.3 Some Positive Development towards Harmonization 147

10.4 Convergence between Certification Systems 147
10.5 Convergence is Happening 148
10.6 Scope for an Omnibus Certification Scheme 149
10.7 Reforms Needed 151
10.8 Conclusions 152

**11. SFM, Forest Certification, and Climate Change Regulations 155**

11.1 Climate Change Imperatives 155
11.2 SFM and Climate Change 156
11.3 Forest Certification and Climate Change 157
11.4 Linkages Expected between Various Climate Change Mechanisms through Risk Management 158
11.5 Similarity between SFM and Carbon Forestry Credit Projects: mechanics of registration and accreditation 159
11.6 Divergence in Objectives and Goals of Carbon Regulatory Mechanism, SFM, and Forest Certification 160
11.7 Common Future of Carbon Regulatory Mechanism (Kyoto Protocol), Certification of Forests, and SFM 161
11.8 Conclusions 162

**12. Developing Certification of Forests in India: rationale and options 163**

12.1 Basics of Certification of Forests in India 164
12.2 National Initiative and Way Forward for Certification in India 165
12.3 SFM and Certification in India 170
12.4 Shape of SFM and Forest Certification: roadmap for evolution 172
12.5 Conclusions 173

**13. Challenges of Certification** **175**

13.1 Important Factors Limiting the Spread of Certification 175

13.2 Challenge of Certification: Indian perspective 179

13.3 Conclusions 183

**14. Future of Forest Certification** **185**

14.1 Bottom Line with Regard to Certification 185

14.2 Opportunities and Limitations of Certification 186

14.3 Future of Certification of Forests 187

14.4 Road Ahead for Certification of Forests 190

14.5 Need for Harmonization 191

14.6 Future Threats to Certification of Forests 192

14.7 Enhancing the Efforts of Sustainable Forest Management and Forest Certification 192

14.8 New Drivers Promoting Certification of Forests 194

14.9 Conclusions 201

*Annexes* *203*

*Bibliography* *231*

*Glossary* *245*

*Index* *255*

*About the Authors* *263*

# Abbreviations

| | |
|---|---|
| ANSI | American National Standards Institute |
| ASI | Accreditation Services International |
| ASEAN | Association of Southeast Asian Nations |
| ATFS | American Tree Farm System |
| ATIBT | International Technical Association for Tropical Timber |
| ATO | African Timber Organization |
| BILT | Ballarpur Industries Ltd |
| C&I | criteria and indicators |
| CAR | corrective action request |
| CDM | clean development mechanism |
| CEF | Certification Española Forestal |
| CEPI | Confederation of European Paper Industries |
| Cerflor | Brazilian National Forest Certification Program |
| CERTFOR | Chilean Forest Certification System |
| CFP | certified forest product |
| CIFOR | Center for International Forestry Research |
| CIS | Commonwealth of Independent States |
| CITES | Convention on International Trade in Endangered Species |
| CNCC | Council of the National Certification Center, Czech |
| CoC | chain of custody |
| CSA | Canadian Standards Association |
| DLH | Dalhoff Larsen and Horneman AS |
| EC | European Commission |
| ECE | Economic Commission for Europe |
| ECOSOC | Economic and Social Council (of UN) |
| EFI | European Forest Institute |
| EMAS | Eco-management and Audit Scheme |

| | |
|---|---|
| EMS | environmental management system |
| ENGO | Environmental non-governmental organization |
| EU | European Union |
| FAO | Food and Agriculture Organization |
| FFCC | Finnish Forest Certification Council |
| FLEGT | Forest Law Enforcement, Governance and Trade |
| FMU | forest management unit |
| FPAC | Forest Products Association of Canada |
| FSC | Forest Stewardship Council |
| GBI | Green Building Initiative |
| GEF | Global Environment Facility |
| GFTN | Global Forest and Trade Network |
| GHG | greenhouse gas |
| GICIA | Green Initiatives Certification and Inspection Agency |
| GMO | genetically modified organism |
| GoI | Government of India |
| HCVF | high conservation value forest |
| IAF | International Accreditation Forum |
| IEC | International Electrotechnical Commission |
| IFF | Intergovernmental Forum on Forests |
| IIFM | Indian Institute of Forest Management |
| IFIR | International Forest Industry Roundtable |
| IIED | International Institute for Environment and Development |
| IPF | Intergovernmental Panel on Forests |
| ISO | International Organization for Standardization |
| ITC PSPD | ITC Paperboards and Specialty Papers Division |
| ITTA | International Tropical Timber Agreement |
| ITTC | International Tropical Timber Council |
| ITTO | International Tropical Timber Organization |
| JCP | Joint Certification Program |
| KP, K/P | Kyoto Protocol |
| LEED | Leadership in Energy and Environmental Design |
| LEI | Indonesian Ecolabeling Institute |

| | |
|---|---|
| LFMU | local forest management unit |
| MC&I | Malaysian criteria and indicators |
| MCPFE | Ministerial Conference on the Protection of Forests in Europe |
| MEA | Millennium Ecosystem Assessment |
| MoEF | Ministry of Environment and Forests |
| MRA | Mutual Recognition Agreement |
| MTCC | Malaysian Timber Certification Council |
| NA | not available |
| NGB | national governing body |
| NGO | non-governmental organization |
| NSC | National Steering Committee |
| NWG | National Working Group |
| NTFP | non-timber forest products |
| OECD | Organization for Economic Co-operation and Development |
| OFI | Oxford Forestry Institute |
| OLB | Origin and Legality of Timber |
| P&C | principles and criteria |
| PAFC | Pan-African Forest Certification Scheme |
| PCI | Principles, Criteria, and Indicators |
| PEFC | Programme for the Endorsement of Forest Certification |
| PEOLG | Pan-European Operational Level Guidelines |
| PPM | process and production method |
| QWEB | Quebec Wood Export Bureau |
| REDD | Reduced Emissions from Deforestation and Forest Degradation |
| REIT | real estate investment trust |
| RIL | reduced impact logging |
| SA | Soil Association |
| SCC | Standards Council of Canada |
| SFI | Sustainable Forestry Initiative |
| SFM | sustainable forest management |
| SIS | Swedish Standards Institute |

| | |
|---|---|
| SME | Small to Medium-sized Enterprise |
| TBT | Technical Barriers to Trade |
| TLTV | Timber Legality and Traceability Verification |
| TRADA | Timber Research and Development Association |
| UKWAS | UK Woodland Assurance Standard |
| UN | United Nations |
| UNCED | United Nations Conference on Environment and Development |
| UNCHE | United Nations Conference on the Human Environment |
| UNCSD | United Nations Commission on Sustainable Development |
| UNECE | United Nations Economic Commission for Europe |
| UNFCCC | United Nations Framework Convention on Climate Change |
| UNFF | United Nations Forum on Forests |
| USDA | United States Department of Agriculture |
| USGBC | United States Green Building Council |
| WB | World Bank |
| WCED | World Commission on Environment and Development |
| WFC | World Forestry Congress |
| WTO | World Trade Organization |
| WWF | World Wide Fund for Nature |

# 1
CHAPTER

# Sustainability and Forestry

## 1.1 CONCEPT

The concept of sustainability has been and can further be used as a mechanism for determining the thresholds of permissible development to fulfill the present and future needs of all communities in the society (given a better current understanding of the concept).

Many efforts in the past to define sustainable development (including sustainable forest management) have always met with confusion and argument. Any attempt to define "sustainability" in a sentence or a paragraph or even in several pages can be limiting, given the complexity of the issue and the volume of studies by experts on this topic.[1]

The concept of "sustainability" found roots in the United Nations' 1987 Brundtland Commission Report, *Our Common Future,* and even earlier in the 1980s in the World Conservation Strategy. Starting from a "pure" ecological concept in the 1970s, World Conservation Strategy, it transformed quickly into a more comprehensive socio-economic approach (Lutteken and Hagedorn 1999). *The Brundtland Report* of the World Commission on Environment and Development (UN 1987) defines sustainability as follows.

"...development that meets the needs of the present without compromising the ability of future generations to meet their own needs".

"...in essence, sustainable development is a process of change in which the exploitation of resources, the direction of investments, the orientation of technological development and institutional change are

[1] Details available at <www.forestry.about.com/od/certification/Forest_Certification.htm>

all in harmony and enhance both current and future potential to meet human needs and aspirations".

There are two main characteristics associated with the term "sustainability" (Lutteken and Hagedorn 1999).

1. "It is a people-centered approach which aims to improve the quality of human life and it is conservation-based in nature in that it is conditioned by the need to respect nature's ability to provide resources and life-support services. In this perspective, sustainable development attempts to improve the quality of human life within the carrying capacity of supporting ecosystems".
2. "It is a normative concept that embodies standards of judgment and behaviour to be respected as the human community 'the society' seeks to satisfy its needs of survival and well-being".

## 1.2 EMERGENCE OF THE CONCEPT OF SUSTAINABILITY IN FOREST MANAGEMENT

### 1.2.1 Origins of Sustainability in Forest Management

Sustainability in forest management began as a biological concept, which later embraced socio-economic aspects as well. Early forest managers had similar concerns and thus developed an understanding of natural forestry productivity and strived to enhance it through silvicultural science to maintain a continuous supply of wood, wildlife, and other products and services for human use and consumption. This concept and action (a limited extent of sustainability) was fundamentally driven by the desire to avoid the social and economic disruption associated with shortages of timber, whether for local use or as the basis for a community or export economy. In this approach, forest products have clearly held the potential of being a perpetually renewable resource and foresters undertook the responsibility of making this happen (Sample 2004).

The origins of sustained-yield forest management can be found in late medieval Europe also (Heske 1938). The lack of well-developed systems for transportation and communication at that time resulted in a system of small, independent political units with high customs barriers that prevented any significant degree of regional trade (Waggener 1977). Local consumption was almost entirely dependent on local production and communities had to be largely self-sufficient.

Thus there was a distinct possibility of exhausting local timber resources unless collective use was strictly controlled and that forced the production and consumption of forest products to become highly regulated. This applied not only to the cutting of timber and wood for fuel but to the gathering of leaf litter and grazing of livestock, both of which were understood to affect long-term soil productivity in forests. It has been argued (Adams 1993) that the concepts of secured land tenure for private property owners, mutual consent under common law, and government intervention in free markets to protect broader public interest—principles basic to the development of constitutional democracies—had their origins in communities such as these, seeking to avoid a "tragedy of the commons" (Hardin 1968). Perhaps because of the opportunities it facilitated for stable employment and income in rural communities, the approach to sustained-yield forestry persisted long after improved transportation and communication systems had reduced the need for local self-sufficiency and turned wood into a widely traded economic commodity (Sample 2004). Box 1.1 elaborates the origin of the term "sustainable forest".

**Box 1.1** Origin of sustained-yield forestry or "sustainable forest"

The words "sustainable forest" or sustained yield became prominent among the foresters in Europe during the eighteenth and nineteenth centuries. At that time, much of Europe was being deforested, and foresters became increasingly concerned since wood was one of the driving forces of the European economy. Wood was used for heating homes, building houses and factories, and making furniture and other articles. Thus the forests that provided wood were central to economic activity. The idea of sustainability became popular and was later brought to the USA and popularized by foresters such as Fernow, Pinchot, and Schenck among others.

*Source* <www.forestry.about.com/od/certification/Forest_Certification.htm>

It was in this context that the concept of "regulated forest" came into being. A regulated forest is one that manages to yield regular, periodic, and sustainable harvests of timber and other products and services. The objective of sustained-yield management by itself does not indicate a single specific harvest level since a forest can be sustained at a range of different management intensities (Sample 2004).

Increase in sustainable forest management increase during the 1980s (Box 1.2).

**Box 1.2** Clear felling of tropical forests: rise of sustainable forest management

During the 1980s, large areas of tropical forests were cleared for cultivation or plantation of exotic tree species for timber production. Some forests were also simply clear felled, and the lands were abandoned. This huge land use conversion raised a lot of concern and became a trigger to promote the practice of sustainable forest management and certification worldwide.

*Source* Perera and Vlosky (2006)

## 1.3 LANDMARK EVENTS IN SUSTAINABLE FORESTRY

To address the apparent conflict between the interests of economic development and the interests of environmental protection, the United Nations in 1983 appointed an international commission to propose strategies for improving human well-being in the short term without threatening the local and global environment in the long term. Norwegian Prime Minister Gro Harlem Brundtland chaired the commission and came out with the report *Our Common Future*, widely known as *The Brundtland Report* (UN 1987). *The Brundtland Report* helped popularize the term "sustainable development" and defined it as "development that meets the needs of the present without compromising the ability of future generations to meet their own needs". This landmark report helped trigger a wide range of actions, including the United Nations Conference on Environment and Development (UNCED) or the "Earth Summit" in 1992 in Rio de Janeiro.

The single most important outcome of the Earth Summit in Rio in terms of defining sustainable forestry was a set of "forest principles", with linkages to several other new international conventions on biodiversity, climate change, and desertification. The forest principles, in turn, served as the basis for several subsequent important developments in international forest policy.

Subsequently, in Montreal in 1993, many of the world's industrialized nations (countries with temperate and boreal forests, including the USA) developed a comprehensive set of *Criteria and Indicators for the Conservation and Sustainable Management of Temperate and Boreal Forests*, which, following their ratification in Santiago in 1995, became what is now commonly referred to as the *Montreal Process Criteria and Indicators* or simply the C&I (Castañeda, Palmberg-Lerche, and Vuorinen 2001). The framework was visualized to have an important role in measuring the progress of each country towards sustainable forest management. The framework was based on seven criteria and 67 indicators, known collectively as the *Montreal Process Criteria and Indicators for the Conservation and Sustainable Management of Temperate and Boreal Forests* (Washburn and Block 2001). Of the seven criteria, the first criterion dealt with the conservation of biodiversity. This process and the discussions also emphasized the social expectation that forests be managed in a way that ensures sustainability of a variety of resources. The seven criteria are as follows (Castañeda, Palmberg-Lerche, and Vuorinen 2001).

1. Conservation of biological diversity.
2. Maintenance of productive capacity of forest ecosystems.
3. Maintenance of forest ecosystem health and vitality.
4. Conservation and maintenance of soil and water resources.
5. Maintenance of forest contribution to global carbon cycles.
6. Maintenance and enhancement of long-term multiple socio-economic benefits to meet the needs of societies.
7. Legal, institutional, and economic framework for forest conservation and sustainable management.

Later at a conference in Helsinki in 1994, most of the tropical timber consumer nations made a joint commitment to maintain, or achieve, sustainable management of their forests by 2000. The details are given in Section 1.6.

Box 1.3 discusses the criteria and indicators approach for sustainable forest management.

**Box 1.3** "Criteria and indicators" approach for sustainable forest management: a critique

Some people believe that forest assessments based on the "criteria and indicators" are not action oriented. The criteria and indicators simply provide a basis for assessing current conditions and trends; they do not, in themselves, provide a basis for the evaluation of adequacy of current efforts to conserve forests, nor are they a basis for identified specific actions to correct any perceived shortcomings in current efforts. However, there is a parallel international process for accomplishing this.

*Source* Sample (2004)

In the years following the Earth Summit of 1992, an array of developed and developing nations requested the United Nations to serve as convener of the Intergovernmental Panel on Forests (IPF), whose purpose was to develop "proposals for action"—specific activities aimed at exhorting more nations towards achieving sustainable forest management. The IPF and its successor, the Intergovernmental Forum on Forests (IFF), which represented a broader array of participants, eventually produced a set of 270 "proposals for action". Each of the proposals for action can be linked to specific indicators of sustainable forest management (SFM) in the criteria and indicators (Washburn and Block 2001).

In 2000, the Economic and Social Council (ECOSOC) of the United Nations decided to create a new office, the United Nations Forum on Forests (UNFF), with a five-year (2001–05) programme of work to facilitate monitoring, assessment, and reporting of progress in implementing the proposals for action.

So although the criteria and indicators themselves do not require nations to take specific actions to achieve sustainable forest management, the national-level and other assessment bodies utilizing the criteria and indicators do help to promote a common understanding of the current status of a nation's forests. Indirectly, this helps to stimulate timely and effective action, so that when the next assessment is conducted in subsequent years, substantial progress towards sustainable forestry goals that countries have set for themselves may be documented (Sample 2004).

Box 1.4 discusses the how indigenous people are affected by sustainability approach in forest management.

**Box 1.4** Sustainability approach in forest management helps indigenous people

One such forest (managed on the principles of sustainability) belongs to the Menominee Indian Tribe of Wisconsin, USA. Over the last 140 years, the Menominee tribe has removed two billion board-feet of lumber. Even so, the volume of saw timber now is greater than that in 1854 when the Wolf River Treaty created the present Menominee Reservation. They are a prime example that human benefits can be provided by a sustainably managed forest.

*Source* <www.forestry.about.com/od/certification/Forest_Certification.htm>

## 1.4 SUSTAINABILITY: CONCEPT IN FORESTRY

There have been various attempts to define sustainable forest management, many of which have tried to capture it with differing types of criteria and indicators. The International Tropical Timber Organization describes it as "the process of managing forests to achieve one or more clearly specified objectives of management with regard to the production of a continuous flow of desired forest products and services without undue reduction of its inherent values and future productivity and without undue undesirable effects on the physical and social environment" (ITTO 2008).

Thus SFM is a series of guiding principles, developed by the international community and adapted and adopted by a number of countries, to secure the role and importance of forestry in national and international environmental, ecological, economic, and commercial terms. SFM is an inherent aim of certification. It is the ultimate goal to which certified forests should aspire, but such a goal is reached only through a period of transition, during which management standards are progressively established and fine-tuned. The explicit aim of certification is to improve the quality of forest management to reach this goal (Upton and Bass 1995).

All attempts for certification of forests have a basis in ensuring sustainability in forestry operations. It has been widely realized that "unsustainable" forest management has been causing "massive soil erosion and biodiversity loss, as well as negatively affecting the economic future of many communities around the globe, especially in the tropics".[2]

Poor forest management practices and indiscriminate logging often lead to biodiversity and habitat loss at the expense of local communities. The World Bank estimates the global market value of losses from illegal logging at over $10 billion and annual losses in government revenues at about $5 billion.

Box 1.5 expounds earlier attempts to practice sustainable forest management.

---

[2] Details available at <www.forestry.about.com/od/certification/Forest_Certification.htm>

**Box 1.5** Earlier attempts to practice sustainable forest management

The American Forest and Paper Association (AF&PA) represents the North American industrial attempt to deal with sustainability in forestry. They made an early and alternative approach, which may be considered a bit more realistic in the North American context.

Sustainable forestry principles were developed by AF&PA under the banner of the Sustained Forestry Initiative (SFI). Its objective was to achieve a much broader practice of sustainable forestry throughout the USA without higher cost for the consumer. The AF&PA membership is primarily the forest industry, and its concern is basically economic.

*Source* <www.forestry.about.com/od/certification/Forest_Certification.htm>

Taking the example of the industrial forestry sector, industry associations like the American Forest and Paper Association (AF&PA) are of the view that "sustainable forestry" is a dynamic concept that will evolve with experience and that new knowledge provided through research will be useful in its evolution. The AF&PA considered its "principles and implementation guidelines" another step in the evolution of industrial forestry practices in the USA. Through this step, the AF&PA members seek to meet the needs of humanity for essential wood and paper products while protecting and enhancing other forest resource values.[3]

## 1.5 SUSTAINABLE FORESTRY: VARIOUS FACETS

SFM is a regime that integrates and balances social, economic, ecological, cultural, and spiritual needs of present and future generations. Essential aspects of SFM include the following (Figure 1.1).

- *Economic sustainability* For economic sustainability, benefits to concerned group(s) should exceed the costs incurred and some form of equivalent capital should be handed down from one generation to the next. The capacity of forests to attract investment and support economically viable uses in the present and future should be undiminished. The forest is not used beyond its long-term capacity for production of wood and non-wood forest products.

[3] Details available at <www.forestry.about.com/od/certification/Forest_Certification.htm>

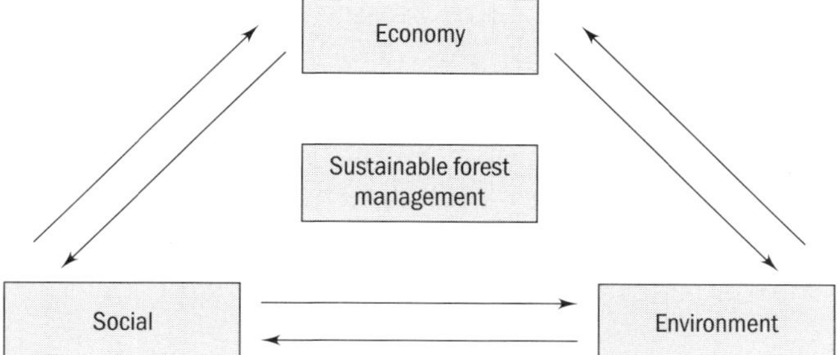


**Figure 1.1** Three pillars of sustainable forest management
**Source** <www.benjapan.org/iceab10/8.pdf>

- *Social sustainability* The relationship between development and social norms is reflected in social sustainability. An activity is socially sustainable if it conforms to social norms or does not stretch them beyond a community's tolerance for change (Upton and Bass 1995). It includes the following aspects.
    - The rights of indigenous peoples and local communities are respected and protected.
    - Forest workers are healthy, safe, and their rights are protected (for example, freedom of association, right to bargain, freedom from child and forced labour, equal remuneration, and non-discrimination).
    - Local communities, including indigenous people, benefit economically from forest management.
    - Sites of religious, spiritual, archaeological, historic, as well as aesthetic and recreational values are preserved.
- *Environmental sustainability* For environmental sustainability, an ecosystem should be able to support healthy organisms, while maintaining its productivity, adaptability, and capability for renewal. It requires that forest management respects and builds on natural processes (Upton and Bass 1995). It means that forest use protects biodiversity (ecosystems, species, genes, and ecological processes) and the capacity to maintain ecosystem processes and services such as watershed protection, pollination, protection against mudslides, aesthetic beauty, and carbon storage.

There are various approaches, positions, standards, and definitions of SFM and specific management measures it requires. There are also various methods to measure progress towards sustainable forest management. Depending on the way proponents understand the concept and management objectives, SFM standards for the same forest could be different. Thus regional standards for SFM can legitimately be somewhat different from one another, reflecting differences in forest types, legal frameworks, social conditions, and other factors. However, mainstream standards for SFM differ on the following issues.[4]

- *Clear-cutting* SFM standards, including Canadian Standards Association (CSA), Forest Stewardship Council (FSC), Programme for the Endorsement of Forest Certification (PEFC), and Sustained Forestry Initiative (SFI), recognize clear-cutting as consistent with SFM in the right forest ecosystems. Clear-cutting can accomplish the following.
  - It mimics some of the natural disturbance dynamics of forests (for example, fire, wind blow downs, and insects).
  - In some ecosystems, it allows regeneration and rapid growth of certain tree species.
  - It costs less, making forestry more economically viable.
  - It provides safer working conditions for loggers.
  - However, all SFM standards recognize that there is no single harvesting method suitable for all forest ecosystems.
- *Plantations* Plantations can focus production on smaller but more intensively managed areas. All SFM standards recognize plantations as being consistent with SFM under certain conditions, which may include considerations based on ecological systems of the place.
- *Chemicals* Most standards allow controlled and appropriate use of chemicals (pesticides and fertilizers). But some standards prohibit the use of chemicals.
- *Genetically modified organisms* Some standards strictly prohibit the use of genetically modified organisms while others allow the use if and when legally available. At least 24 tree species have been known to be the subject of transgenic research.

[4] World Resource Institute and World Business Council for Sustainable Development; details available at <www.sustainableforestprods.org/node/>

## 1.6 GOAL OF SUSTAINABLE FOREST MANAGEMENT

- In making the transition to SFM in any given locality, the following aspects also need to be included (Upton and Bass 1995).
- Achieving environmental objectives, such as maintaining biodiversity, watershed quality, and climate regulation.
- Achieving economic objectives, such as maintaining timber yield and the forest's capital value.
- Achieving social objectives, such as meeting livelihood needs and maintaining the cultures and knowledge systems of forest-dependent people.
- Balancing today's needs with those of future generations.
- Integrating the above objectives where possible; exploiting "win-win" possibilities
- Making informed trade-offs between objectives where integration is not possible; cost–benefit analysis and environmental impact assessment can help to achieve maximum positive impact and minimum negative impact, both on-site and off-site.
- Balancing the different levels at which needs and demands are expressed, from the forest compartment to neighbouring communities and land uses, from the nation to the globe; at each level, integration and trade-off possibilities will be different, and consensus building and negotiation can help.
- Allowing for complexities, and many uncertainties; we will not always know the possible outcomes.
- Adopting procedures for continuous improvement, which requires an emphasis on monitoring and learning.
- Ensuring participation of all stakeholder groups in decision-making and transparency of decisions.
- Providing good information access to all stakeholder groups with legitimate interests.
- Gaining the commitment of "losers" of each decision, as well as "winners".
- Ensuring that stakeholders are held accountable for their forest activities, given that these will have an impact on other stakeholders; this may require independent verification, such as certification.
- Long-term, policy-level support and financial stability for SFM.

Sustainability depends on the specific relationships of forest management with the surrounding environment and society. Few criteria of sustainability will be universal. Precise definitions of sustainability need to be locally negotiated and revised in accordance with the demands of technology and society and other changes in condition (Upton and Bass 1995).

The relative importance of effects and objectives will always differ according to the situation. For example, achieving sustainability for plantations may require a focus on the impact on soil and water regimes, work safety, and employment. Focus should also be on equity impacts in terms of changed access to food, water, and firewood, and impacts on incomes. A sustainable plantation would be likely to adopt priorities based on the following (Upton and Bass 1995).

- Profitable and efficient production, with an emphasis on improved whole-forest, site-specific management and which uses a systematic approach to conserve soil, water, energy, and biological resources.
- A more thorough incorporation of natural processes, absence of which harms the environment or health of forest operatives and a more targeted and efficient use of the remaining inputs used.
- A greater productive use of the biological and genetic potential of timber and animal species.
- An improvement in the match between tree species and the productive potential and physical limitations of forested lands to ensure the long-term sustainability of current production levels.

In contrast, sustainability of tropical rainforest management would require additional emphasis on significant biodiversity, customary access, and the availability of non-wood forest products (Upton and Bass 1995).

It is perhaps clear from above that achieving the goal of SFM is a multidimensional balancing act. In today's world, new procedures and initiatives are required to support the process of achieving SFM. Forest certification has been designed ultimately to lead to SFM (Upton and Bass 1995).

## 1.7 GENESIS OF SUSTAINABLE FOREST MANAGEMENT

As discussed earlier, a series of international and national initiatives led to the development of SFM principles and criteria. Some of the agreed-upon national and international political commitments that led to the evolution of SFM are discussed in the subsequent sections.

### 1.7.1 Earth Summit of Rio

The UNCED, also popularly known as the Earth Summit, took place in Rio de Janeiro, Brazil, during 2–14 June 1992. It was held 20 years after the United Nations Conference on the Human Environment (UNCHE) in Stockholm, Sweden. Government officials from 178 countries and between 20 000 and 30 000 individuals from governments, non-governmental organizations (NGOs), and the media participated in this event to discuss solutions to global problems such as poverty, war, and the growing gap between industrialized and developing countries. The central focus of discussion was how to incorporate these concerns in global environmental systems and the paradigm of sustainable development. This ultimately broadened the scope of global environmental diplomacy by adopting the notion of sustainable development, which had been advocated five years earlier by the World Commission on Environment and Development as one of its key policy frameworks.

The outcome of this summit was the Rio Declaration. This enunciated principles of environment and development in the form of *Agenda 21*, a statement of principles for the sustainable management of forests, which were adopted by consensus (without vote).

The institutional innovation resulting from the conference included an agreement on the operating rules for the Global Environmental Facility (GEF), United Nations Convention on Biological Diversity (UNCBD), and establishment of the United Nations Commission on Sustainable Development (UNCSD) on the basis of *Agenda 21* recommendations. The United Nations Framework Convention on Climate Change (UNFCCC) and UNCBD were the products of independent but concurrent negotiating processes that were opened for signatures at the UNCED (Cleveland, Kubiszewski, and Miller 2007). The Rio Declaration on Environment and Development is a set of 27 legally non-binding principles designed to commit governments to ensure environmental protection and responsible development and intended to be an Environmental Bill of Rights, defining the rights of people to development and their responsibilities to safeguard the common environment. It established the "precautionary principle" and the principle of "common but differentiated responsibilities". The declaration recognizes that the only way to have long-term social and economic progress is to link it with environmental protection and to establish equitable global partnerships between governments and key actors of civil society and the business sector.

### 1.7.2 *Agenda 21*

*Agenda 21*, the international plan of action for sustainable development, outlines key policies for achieving sustainable development that meets the needs of the poor and recognizes the limits of development to meet global needs. *Agenda 21* can be termed a blueprint for sustainability and forms the basis for sustainable development strategies. It attempts to define a balance between production, consumption, population, development, and the Earth's life-supporting capacity. It addresses poverty, excessive consumption, health, education, cities, agriculture, food and natural resource management, and several more subjects (Cleveland, Kubiszewski, and Miller 2007).

### 1.7.3 Intergovernmental Initiatives to Address SFM

Some of the important intergovernmental initiatives to address SFM are as follows.

- Helsinki Process (boreal forests)
- Montreal Process (temperate forests)
- Tarapoto Proposal (tropical forests)
- Pan-European Forest Process

Details of other important processes are given in Table 1.1. Each of these processes defined the principles of SFM in terms broadly applicable to major forest types as well as proposed criteria and indicators by which SFM can be monitored, assessed, and promoted at national and international levels.

Another important feature of SFM standards is that they must be specific for the national or local conditions prevailing in different types of forests with regard to their ecosystem functions and socio-economic values. The diversity of forest management makes uniform standards difficult to apply; therefore, only general requirements for SFM may be agreed upon at an international level. Such an agreement exists in various sets of regional criteria and indicators for SFM, among which the International Tropical Timber Organization's criteria and indicators have played a pioneering role (Castañeda 2000). These criteria and indicators were originally developed for monitoring forest

**Table 1.1** Countries participating in major international processes on criteria and indicators for achieving SFM

| *Process* | *Number of criteria* | *Number of indicators* | *Scope* | *Place of adoption* | *Date of implementation* | *Number of countries** |
|---|---|---|---|---|---|---|
| ITTO initiatives in criteria and indicators# | 7 | 66 | National and FMU levels in humid tropical forests of member tropical countries | Yokohama, Japan | March 1992 | 12 |
| Dry-zone Africa process | 7 | 47 | National level | Nairobi, Kenya | November 1995 | 28 |
| Pan-European Forest Process | 6 | 27 (quantitative) 101 (descriptive) | Boreal, temperate, and Mediterranean-type forests in Europe, regional, and national levels | Helsinki, Finland; Lisbon, Portugal | June 1993, June 1998 | 37 |
| Montreal Process | 7 (non-legally | 67 | Temperate and boreal forests in countries outside Europe—national level | Santiago, Chile | February 1995 | 12 |
| Tarapoto Proposal | 1 (global) 7 (national) 4 (FMU) | 7 (global) 47 (national) 22 (FMU) | Sponsored by the Amazon Cooperation Treaty | Tarapoto, Peru | February 1995 | 8 |
| Near East Process | 7 | 65 | Regional and national level | Cairo, Egypt | October 1996 | 30 |

*Contd...*

**Table 1.1** *Contd...*

| *Process* | *Number of criteria* | *Number of indicators* | *Scope* | *Place of adoption* | *Date of implementation* | *Number of countries** |
|---|---|---|---|---|---|---|
| Lepaterique Process of Central America | 4 (regional)<br>8 (national) | 40 (regional)<br>53 (national) | Forest management level criteria and indicators and identified at subregional meetings and national seminars | Tegucigalpa, Honduras | January 1997 | 7 |
| African Timber Organization | 28 | 60 | ATO member countries; regional and national levels; also identified five principles and two sub-principles | Libreville, Gabon | January 1993 | 13 |
| Regional Initiative for Dry Forests in Asia | 8 | 49 | Dry forests in Asia; national level | Bhopal, India | December 1999 | 9 |

FMU – forest management unit; ITTO – International Tropical Timber Organization

*The total number of countries participating in the international processes adds up to 156 because many countries are members of more than one process.

#Out of the 55 members of the International Tropical Timber Organization (ITTO), only 12 listed countries participate in the ITTO's criteria and indicators process. The rest of the ITTO members participate in other processes. India, Myanmar, and Thailand are also members of the regional initiative for dry forests in Asia.

**Source** Castañeda (2000)

management (ex post), but they have started to affect the setting of objectives and performance requirements (ex ante) (Rametsteiner and Simula 2001). Examples of the latter include the Pan-European Operational Level Guidelines (PEOLG), which are used by the PEFC as a reference basis and are related to the pan-European criteria and indicators for SFM (Atyi and Simula 2002). These issues are also important in an Indian context and are discussed in Chapter 12. An initiative of great significance for tropical forests is the International Tropical Timber Agreement (ITTA), which was set up in 1983 as a forum for consultation and cooperation among tropical timber producer and consumer countries.

## 1.8 BASIC REQUIREMENTS FOR ACHIEVING SUSTAINABLE FOREST MANAGEMENT

The basic structural requirement for any framework for SFM to happen includes the following (Costa, Salmi, Simula, *et al.* 1999).

- *Policy* An effective and appropriate policy and regulatory framework, that is, a clearly defined forest policy, integrated with other sectoral policies, especially with regard to land use allocation issues.
- *Legal* Clear legal framework, particularly with regard to land tenure rights and allocation, including effective enforcement procedures and structures.
- *Political* Stable political environment with transparent, participatory, and democratic decision-making processes.
- *Institutional* Institutions that function well, that is, streamlined public–private sector interface, on-going capacity building to increase efficiency of public bodies, and effective organization of private sector actors to represent and promote their interests.
- *Market* Adequate market infrastructure, including appropriate level of government intervention (but usually low).
- *Fiscal* Well-conceived objective led tax/subsidy regime (incentives/ disincentives)
- *Macroeconomic* Stable macroeconomic horizon conducive to investors.

The above factors have a positive effect on SFM, and the absence of any of these would have a negative bearing on the overall success of SFM.

The categorization of stages and the related operational requirements for SFM are summarized in Table 1.2.

The aforementioned ingredients reflect in seven globally applicable criteria for SFM identified by intergovernmental processes for criteria and indicators (Rametsteiner and Simula 2001).

- Extent of forest resources
- Forest health and vitality

**Table 1.2** Operational requirement of SFM in different stages

| *Categorization of stage* | *Requirements* | *Possible options for meeting requirements* |
|---|---|---|
| Pre-operational stage | Research, such as multiple-objective forest inventories | • Dissemination mechanisms (such as extension services) for existing information and research methodologies.<br>• Collaboration with research bodies/non-governmental organizations. |
| | Additional capital investments, for example, in comprehensive management plans | • Incremental cost support for transitional phase to SFM, particularly for global benefits of forests.<br>• Packaging of different financial investment streams (concessionary/commercial); credit facilities for micro-investors; investment guarantee facilities. |
| | Technical assistance | • Extension services; identification and facilitation of appropriate knowledge and skills resources.<br>• Technology transfer. |
| | Participation of local stakeholders | • Development of appropriate methodologies for private sector operators. |
| Operational stage | Training and technical assistance | • Extension services; identification and facilitation of appropriate knowledge and skills resources.<br>• Technology transfer. |
| | Stand regeneration and tending | • Ensuring that appropriate silvicultural systems are established and that they are made financially viable. |
| | Marketing and sales assistance | • Brokerage/packaging of project outputs for sale (for example, through public sector forest investment agency); price guarantees for non-commercial investors. |
| Post-operational stage | Monitoring | • Collaboration with local communities; established dispute resolution procedures; secure land tenure arrangements and enforcement regimes. |

Source Costa, Salmi, Simula, *et al.* (1999)

- Productive functions of forests
- Biological diversity
- Protective functions of forest
- Socio-economic benefits and needs
- Legal, policy, and institutional framework

The policy, market, and institutional failures, inadequate tenure, rising populations and their demands, fragmentation of forest estate, and inappropriate infrastructure, technology, and skills are some of the basic causal agents of problems in forests. The transition to SFM primarily involves tackling these problems (Upton and Bass 1995). This requires policy decisions to be made at national and international levels. Annex 1 gives an overview of strategy for sustainable forestry at the national level.

There are a few debatable facts of sustainable forest management (Box 1.6).

**Box 1.6** Debatable facts of sustainable forest management

Sustainable forest management (SFM) still remains a debatable if not a controversial issue. The fact that it has more than 60 definitions in the literature is evidence enough of that. Most of the controversy is over what exactly SFM means and then, whatever it means, how to implement it. The International Tropical Timber Organization's definition of SFM, which is as good as any and better than most, shows that controversy is an inherent characteristic when one attempts to describe it. How much reduction in values and productivity, and how many undesirable physical and social effects, are too much ("undue") and who says so are the types of questions always raised about SFM. Nevertheless, there is a high degree of unanimity about at least three aspects of SFM. First, there is virtually total agreement that SFM is essential and not just desirable, especially for tropical forests. Second, there is nearly complete agreement that SFM will cost more to take care of than the present mix of neglect, abuse, exploitation, management, conversion, and mismanagement that prevails in forestry. Third, there is total, universal, and unanimous agreement that somebody else should pay for the above cost. From the above, it follows that much more is needed to be done about SFM than just keep talking about it. The conundrum then is that, no matter how essential it may be to have SFM in the tropics, it is not very easy to put into practice. This conclusion of helplessness or despair is not confined to tropical forests but even to forests in general. A collective human solution to this global or national problem is required, and this has huge economic implications. Is there really any easy and comprehensive way?

*Source* Leslie (2006)

## 1.9 EFFORTS TO ENSURE SUSTAINABILITY IN INDIAN FORESTS

### 1.9.1 Scientific Forest Management in India: Historical Evidence

Scientific forestry in India began in 1842 with the initiatives taken by Conolly, Collector of Malabar, under the British. Forest management marked a break from a long existing practice. In 1864, Dietrich Brandis, a German forester, was appointed as the first Inspector General of forests in India. This was followed by the 1865 Forest Act and its subsequent revisions. The concept of working plans and management units and divisions and circles emerged for the scientific management of state-owned forests on a sustainable basis during this time (Rao, Sagaraiya, Bhadran, *et al.* 1961).

Earlier in 1797, German forester George Harting came up with the concept of "sustained-yield management" by which he meant that to ensure continuous wood supply over generations, harvest should not exceed growth. This idea formed the backbone of modern forest management in Europe and North America and many other parts of the globe and also in India. The sustained-yield principle has always been the keystone in the working plan prescriptions in India and was essentially for the harvesting of timber. It essentially meant that the cumulative tree increment should only be harvested leaving the capital intact.

Ancient Indian texts of the Mauryan period (fourth–third century BC) are a testimony to the fact that the concept of sound forestry and forest management practices was well developed during the time. Information about forests during this period is available from various descriptions in scriptures. These are *Arthashastra* of Kautilya (321 BC), *Indika* of Megasthenes (305 BC), *Mudrarakshasa* of Visakhadatta, and the inscriptions of Ashoka (273–236 BC) (Dwivedi 1980). The primary document that highlights the scientific approach to aspects of administration, forest policy, and their enforcement is the *Arthashastra* of Kautilya, which is a treatise of the ancient Indian administration of natural resources. As per these records, there was a regular and independent forest department in the administration of King Chandragupta Maurya. The *Arthashastra* also mentions functional types of forest (Prasad 1999; Rangrajan 1987; Rawat 1991). Although on the face of it, these classifications of forests appear to have only a utilitarian purpose, they also have some basis in sustainability. Some of the forest types of those periods were as follows.

- Pashuvan (game forest)
- Mrigvan (deer forest)
- Dravyavan (production forest)
- Hastivan (elephant forest)

During the Mughal period (1526–1700), the emphasis was predominantly on agriculture. Evidence shows that people were encouraged to clear forest tracts for agricultural purposes. Most of the Mughal buildings are examples of high standard stone masonry, and use of timber in these was comparatively less. The Mughals had a utilization and aesthetic approach towards forests without any comprehensive outlook on the problems of forestry, including aspects of propagation, protection, management or improvement. Up to the middle of the nineteenth century, Indian forests were subjected to large-scale exploitation by ship building and railway sleeper production industries without any effort towards conservation or management (Dwivedi 1980).

## 1.10 MODERN EFFORTS TOWARDS SUSTAINABILITY IN INDIAN FORESTRY

The forestry sector in India is among the first few in the world to be managed on the lines of modern scientific management. Establishment of modern and scientific forest management system and process, which started from the middle of the eighteenth century, coincided with the Industrial Revolution in the West. The forests were considered and used as important resources during the pre-independence period, as the demand for raw materials produced from forests increased. They were extensively exploited to expand the railway network, for example, in the shape of wooden sleepers in that period. Forestry was thus production oriented at that time. However, the basic change in perception that forest as a resource cannot be exploited unhindered was brought about by the National Forest Policy of 1952. This tried to shift focus to production forestry and also had an eye on meeting objectives regarding the maintenance of ecological balance and meeting the needs of various other stakeholders in the best way possible.

## 1.11 INITIATIVES TO INTEGRATE ELEMENTS OF SFM IN CURRENT INDIAN FORESTRY

As discussed earlier, the criteria and indicators approach was developed with the idea of a set of specific forestry conditions evolved through various international processes among participating countries. In this initial phase, it was thought to develop a framework for SFM for the Indian context also to establish a benchmark for sustainability according to the prevailing global policy framework. In 1999, a workshop on "Development of National Level Criteria and Indicators for the Sustainable Management of Dry Forests in Asia" (also called the Bhopal India Process) was held at the Indian Institute of Forest Management (IIFM), Bhopal, with support from the Food and Agriculture Organization of the United Nations and the United Nations Environment Programme in collaboration with the ITTO, the US Department of Agriculture's Forest Service, and the IIFM, Bhopal. The outcome and approach is now widely referred to as the "Dry Forest in Asia Process" in which 10 Asian countries jointly developed a regionally applicable set of national-level criteria and indicators relevant for dry forests in the region (FAO 2000). The Asia regional initiative was later endorsed by the "National Task Force on Sustainable Forest Management", appointed by the Ministry of Environment and Forests (MoEF), Government of India, and these have led to a national draft set of criteria and indicators. Thus the Indian initiative of criteria and indicators approach for SFM has been spearheaded by the IIFM, in collaboration with the ITTO and the MoEF (IIFM 2000). The IIFM, through a series of national technical workshops and consultation meetings, has also been a pioneer in sensitizing communities, forest managers, NGOs, and researchers and emphasizing the need for developing a set of criteria and indicators at the national, state, and forest management unit levels. Over a period of time, eight criteria and 51 indicators specific to Indian forestry conditions were evolved after a consultative process involving a large gamut of stakeholders. The Bhopal India Process has helped evolve the national draft standards after a lot of deliberations and field-testing over the years (Rawat, Menaria, Dugaya, *et al.* 2008). But Indian forestry has still to go a long way to achieve sustainability.

# 2
CHAPTER

# Concept and Evolution of Forest Certification

## 2.1 MEANING OF FOREST CERTIFICATION

Forest management certification is the process by which forest management practices are evaluated against a set of standards. The term "certification" is now commonly understood as "independent verification" (Mater 1999; SAF 1999) of conformity to established standards of sustainable forest management, generally done by a third party. It is a tool designed to document and reward specific forest management practices and to assure consumers of forest products that their purchase comes from a forest whose management meets certain standards. Certification of forests or forest products could be on the basis of many other criteria, which are illustrated in Table 2.1. However, in this book, certification of forest normally implies certification with respect to conformance to sustainable forest management.

Certification is thus a mechanism for ensuring that forests are managed following the principles of sustainable forest management. In this process, forest management practices are evaluated and verified against a set of standards by an external certification organization. Thus, consumers can be confident that they are buying wood from sustainably managed forest sources with documented proof of sustainability.[1] This process ensures "sustainability of forests" and induces the market to offer a "green premium", which in turn invigorates the sustainability of forests that are managed under this system. A major incentive for producers of timber to seek certification is the concept of price premiums; this implies that certified goods may be more valued by consumers and can, therefore, achieve higher

[1] Details available at <www.forestry.about.com/od/certification/Forest_Certification.htm>

**Table 2.1** A broad classification of certification of forests, based on different criteria

| *Certification for* | *What it assures the customer* | |
|---|---|---|
| | *Properties* | *Characteristics* |
| Product quality | Quality of timber/NTFP products | • Appearance grading<br>• Stress grading |
| | Mechanical properties of timber | |
| Organic in process | Ensures the consumer that the produce is herbal/organic in nature and does not use harmful chemicals | No chemical, fertilizer, pesticides, used. Only organic manure used. |
| Fair labour enforcement | Assures that no child labour has been used in the process and no exploitative labour practices | |
| Legal sourced | Assures that the timber/NTFP has come from a legal source | |
| Fair trade practices | All the responsible trade practices have been incorporated | |
| Sustainably managed forest | Assurance that the timber came from a forest that has been managed on sustainability principles | |

NTFP – non-timber forest products

market prices.[2] Theoretically, certified forests would be environmentally, socially, and economically better and at the same time able to yield timber products on a sustained basis, while maintaining adequate wildlife habitats and protecting water quality. Forest certification can also promote timber production when managed effectively. This claims to maintain the long-term viability of commercial forests, protects biodiversity, and provides a continuous stream of social and economic benefits. Because timber markets largely dictate forestry practices, forest certification has created economic incentives to achieve and maintain high standards for forest management worldwide.[3]

In certification, it is preferred to use the term "quality forestry" to describe forest management performance considered adequate, that is, basic to the transition to sustainable forest management (Upton and Bass 1995). Box 2.1 elaborates on the meaning of quality forestry.

---

[2] Details available at <www.forestry.about.com/od/certification/Forest_Certification.htm>

[3] Details abailable at <www.forestry.about.com/od/certification/Forest_Certification.htm>

**Box 2.1** Definition of "quality forestry"

Quality forestry is defined as forestry that is environmentally sensitive, socially aware, and economically viable. (This definition of quality forestry is consistent with the Forest Stewardship Council (FSC) principles and used for certification.)

Forestry is defined as being environmentally sensitive when its impact upon the environment is assessed and negative aspects minimized.

Forestry is defined as being socially aware when it recognizes that its activities have an impact both on local people and society at large. Socially aware forestry encourages all stakeholders to enjoy long-term benefits of forest, provides strong incentives to local people to sustain forest resources and adhere to long-term management plans, and provides for an appropriate distribution of costs.

Forestry is defined as being economically viable when forest operations are structured and managed to be sufficiently profitable for ensuring the stability of operations and a genuine commitment to principles of quality forestry. Economic viability does not mean financial profit at the expense of forest resource, ecosystem, affected communities or the society as a whole.

*Source* Upton and Bass (1995)

Certification is supposed to benefit forest owners economically (at least in theory), as certified forest products should yield a higher return on an investment while assuring consumers that their purchase comes from a forest whose management meets certain standards (Washburn and Block 2001). Thus, certification is an important tool used by both the timber industry and consumers. Timber certification schemes offer buyers confirmation that the origins, processing, and distribution of the timber being sold meet an agreed set of standards.[4]

In fact, certification is a process that results in a written statement attesting to the origin of raw wood material, and its status and/or qualifications following validation by an independent third party. It typically includes two main components (Hansen, Fletcher, Cashore, *et al.* 2006).

[4] Details available at <www.forestry.about.com/od/certification/Forest_Certification.htm>

1. *Forest management certification* based on an assessment of forest management against a set of standards reflecting contemporary concepts of sustainability.
2. *Product certification* involves verifying the chain of custody of wood from the certified source to the consumer.

Certification is also a tool to improve forest harvesting practices and forest management around the world. The philosophy is that the certification label will result in an increased market share for forest owners (Vogt, Larson, Gordon, *et al.* 2000).

Certification of timber can be an application of market tools for achieving widespread sustainable forest management. The usual motivation for a producer or supplier is continued access to markets for timber and wood products, which is partially provided by the certification (Hansen, Fletcher, Cashore, *et al.* 2006). Certification has always been envisaged as a market-driven mechanism that promotes sustainable forest management in three main ways (Durst, McKenzie, Brown, *et al.* 2006).

1. By establishing standards for forest practices and management that guarantee a certain level of management performance.
2. By enhancing marketing opportunities for products from sustainably managed forests.
3. By promoting public education about improved forest management for both producers and consumers.

Certification also provides a mechanism for independent validation of sustainability in markets where forest products are subject to consumer resistance on environmental grounds. Although the initial focus of certification was mainly tropical forests, processes have rapidly expanded to encompass all forest types (Durst, McKenzie, Brown, *et al.* 2006).

The focus of a certification programme is at the forest level. In certification, this area is defined as the local forest management unit (LFMU). Usually, the LFMU is managed by a single owner or operator. However, there are instances where this is not the case. The area being certified must have a common management system but may be owned by several individuals grouped together, by a community or even lie within a single administrative boundary within which forestry activities are regulated by a national forest. In addition, certification will always be of a forest area being managed. Certification, therefore, assumes that prior decisions have been made with regard to different land uses and in setting aside areas as national parks for protection and conservation. Certification cannot assess whether or not a particular forest area should not be managed (Upton and Bass 1995).

## 2.2 NEED FOR CERTIFICATION

An important question that all forest managers always ask themselves while considering certification is whether or not to get the forest certified at all. As stated, certification is the process of verifying that forest management meets the requirements of a responsible forest management standard. It would only be useful for someone to conform to the requirements of a standard if there is a reason for doing so. The most common reasons why a forest manager may decide to pursue certification are as follows (Higman, Mayers, Bass Judo, *et al.* 2006).

- Customers are demanding certified products.
- There is the potential to use certification as a means of accessing new markets.
- An investor, donor or insurer is demanding certification as a conditionality on loan, grant or insurance.
- The owners, shareholders or management see certification as a useful tool to achieve management goals or demonstrate best practices to stakeholders.
- Pressure from government as a condition for the allocation of forest management rights.

Each of these reasons for pursuing forest certification will have different implications in choosing which certification scheme to use. If customers are demanding a specific type of certification, there is little point in achieving forest certification under a different scheme.

## 2.3 OBJECTIVES OF CERTIFICATION

The following is a comprehensive list of important objectives of certification (Durst, McKenzie, Brown, *et al.* 2006).

- Control resource management techniques.
- Control resource economically.
- Alleviate poverty.
- Create a system that assures the public that environmental concerns and values have been addressed.
- Manage resources holistically so that healthy environments are maintained.
- Diminish the amount of regulation being imposed on a forestland owner.
- Balance the need to extract resources from the environment while maintaining sustainable ecosystems.
- Control the values of private forestland owners or enable private forestland owners to maintain their values in the face of society's drive to impose its values on them.
- Add credibility to environment claims.
- Reduce risk of environment non-compliance.
- The possibility that certified products at any stage of the value chain can potentially obtain premiums from buyers.
- Maintain or create access to markets that favour certified products.
- Belief that the process of certification results in improved decision-making and profitability; for example, a certified company has reported a $1 million cost savings in its manufacturing process because of the new programmes initiated through certification.
- Provide forest owners with a means to access higher value markets; for customers to reward the achievement of social and environmental outcomes and for project sponsors to ensure outcomes are consistent with social and environmental expectations.

Certification of plantations according to the Forest Stewardship Council (FSC) approach is explained in Box 2.2.

**Box 2.2** Certification of plantations: the FSC approach

The world recognizes the pressure placed on its forests from the ever-increasing demand for timber resources. It is necessary that the responsible use of natural forests should be a complementary strategy to support the conservation of world's forests.

As a general principle, the more intensive a production system is, the higher the conservation efforts should be. Also by working with different realities between forests and plantations, forest certification systems will be able to achieve its mission of promoting responsible forest management in a better way.

However, no certified plantation can expect to replace natural forests. The Forest Stewardship Council (FSC) does not support conversion of natural forests to plantations or other land uses.

After much debate, the FSC membership has recognized that there is a continuum of conditions between natural forests and plantations, varying from very diverse and complex natural systems to simpler man-made plantations. Principle 10 of FSC pertaining to certification says that plantations should be planned and managed in accordance with Principles and Criteria 1–9 and Principle 10 and its Criteria. While plantations can provide an array of social and economic benefits and can contribute to satisfying the world's need for forest products, they should complement the management of, reduce pressures on, and promote the restoration and conservation of natural forests.

In 1995, the FSC membership voted with a 98% majority in favour of adopting *Principle* 10 for plantations, in addition to Principles 1–9, for plantation management. Producing fibre from *FSC-certified* plantations alleviates pressure from natural forests. Also, many *certified* plantations contribute to the restoration of natural forests by either allowing or actively promoting regeneration of natural forests in sensitive areas such as riparian zones or by creating a buffer around *natural forest* areas. This can be seen in many cases in different parts of the world where plantation type operations have significant areas of natural forests within their boundaries.

FSC is currently reviewing the implementation of the *FSC Principles and Criteria* in plantations. Many FSC members and stakeholders are constructively engaging with FSC in plantations review. In a few instances, however, criticisms have been made to FSC through campaigns.

*Source* <www.fsc.org/plantations.html>

## 2.4 EMERGENCE OF FOREST CERTIFICATION THROUGH HISTORY

Labelling of wood products with a mark of quality can be traced back in Europe to a French royal decree of 1637, which stipulated that members of the guild of cabinet makers had to mark the furniture they made (Pradère 1989).

The genesis of the modern concept of certification can be attributed to society's concern for the social and environmental significance of forests, furthered by the increased environmental awareness in the 1960s and 1970s (Granholm, Vahanen, and Sahlberg 1996; Hansen 1997). Because of high societal awareness and concern towards the increased rate of deforestation in tropical forests, the concept of forest certification emerged in the 1980s as a possible tool to control deforestation and enhance the social and environmental importance of tropical forests in particular.

The concept of modern forest certification evolved in the late 1960s and the early 1970s as a result of environmental awareness and the changes in how society perceived its environment (especially in developed countries). In the 1980s, the high awareness of the loss of tropical forests played a pivotal role in stimulating the development of forest certification as a tool to control deforestation rates (Vogt, Larson, Gordon, *et al.* 2000). The very foundation of forest certification was built on a concern for the destruction of tropical rainforests.

Other forms of labelling of wood emerged in the 1990s under forest certification as a market-based response to address public concerns related to deforestation, mainly in the tropics (Perera and Vlosky 2006).

The use of forest certification as a potential tool to deal with deforestation rates can be traced back to the late 1980s. However, the 1992 United Nations Conference on Environment and Development (UNCED) can be identified as the watershed event (Princen and Finger 1994) for the environment movement as a whole and as the point at which certification became a reality. The UNCED was the first attempt to reach consensus on forest issues. Forestry issues became a part of the international political agenda with ECO 92 (UNCED), when efforts were increased to determine how forests should be managed worldwide on a sustainable basis. After the UNCED, the first set of Principles, Criteria, and Indicators (PCI) were released on how sustainable forest management should be practised and evaluated for certification of forests.

The concept of timber certification emerged in 1990 when a group of stakeholders in California recognized the need for a credible system to identify acceptable sources of forest products (Perera and Vlosky 2006).

The Rainforest Alliance launched its Smart Wood Programme in 1990 as a tool to curb deforestation in the tropics. Their premise was that several independent groups would scientifically evaluate and certify forests that were correctly managed for sustainability and biodiversity.[5]

Later, the FSC, spearheaded by the World Wide Fund and supported by environmental non-governmental organizations (NGOs) and social activists, was established in 1993 to prevent forest deterioration globally and help tropical timber producers to avoid environmental boycotts and meet the demand from importing countries for environmentally sensitive wood products from well-managed forests.

The FSC has a set of principles and criteria for certifying forests. FSC certification is one of the most widely accepted certification schemes, which is mostly favoured by NGOs, but is probably less popular with industry (Bettinger, Boston, Siry, *et. al.* 2009).

The unique concept and part of the certification plan is to use consumer purchasing of forest products as the catalyst for responsible forest management around the world and thus harness market forces to perpetuate a healthy forest with little or no need for regulation.

## 2.5 EMERGENCE OF CERTIFICATION IN AMERICA AND EUROPE

The concept of certification emerged in the USA in 1941 when the American Tree Farm System (ATFS) was created to draw public attention to the active management and investments made in private forests. The ATFS certification was based on a set of forest management principles and required an on-the-ground inspection of forest practices every five years (Bettinger, Boston, Siry, *et al.* 2009).

[5] Details available at <www.forestry.about.com/od/certification/Forest_Certification.htm>

In the 1990s, the American Forest and Paper Association (AF&PA), an industry trade group, initiated its own certification system. The AF&PA added an optional third-party auditing component.

In 1996, the Canadian Standards Association (CSA) developed standards for sustainable forest management, which were updated in 2002. The Sustainable Forestry Initiative (SFI), which was originally an industry self-regulation programme, recognized the CSA system as the functional equivalent of the SFI system (Perera and Vlosky 2006).

The National Forestry Association (NFA) in the USA has developed a system called Green Tag for woodland owners (Perera and Vlosky 2006).

In Europe, forest owners created an alternative to the FSC called the Programme for the Endorsement of Certification scheme in 2003. This was later renamed as the Programme for the Endorsement of Forest Certification (PEFC) scheme.

The International Organization for Standardization (ISO) 14001 Environment Management Standard is not specific to forestry. However, forestry operations can use its environmental management system framework as a foundation to implement a forest management standard (Hansen, Fletcher, Cashore, *et al.* 2006).

## 2.6 CERTIFICATION: AN INITIATIVE TO COMBAT ILLEGAL LOGGING

More than 100 million $m^3$ of timber are cut illegally each year worldwide (Lawson, Sam, and McFaul 2010). Domestic markets in producer countries consume an increasing proportion of illegal timber, much of which come from small-scale artisanal logging. Illegal logging and the associated trade in illegally sourced wood products are important causes of deforestation and forest degradation all over the world. Forest destruction contributes up to 20% of global carbon dioxide emissions. Illegal logging and its associated trades also undermine government revenue, promote corruption, threaten the rule of law, and fund armed conflict (Lawson and McFaul 2010). Forest certification emerged as a private sector-based initiative.

Table 2.2 gives an outline of different types of initiatives evolved to combat illegal wood trade (van Dam and Savenije 2011).

**Table 2.2** Number of initiatives per category, starting date, and type

| *Government initiative side* | *Number* | *Starting date* | *Supply-side measure* | *Demand-measure* |
|---|---|---|---|---|
| Global and multilateral agreements and eclarations | 19 | 1963–2008 | ✓ | ✓ |
| FLEG and FLEGT and similar processes | 8 | 2001–08 | ✓ | |
| Regional bodies and processes | 10 | 1989–2008 | ✓ | |
| Bilateral initiatives | 12 | 2002–09 | ✓ | |
| Free trade agreements | 7 | 1991–2008 | ✓ | ✓ |
| National policies and legislation | 5 | 2005–09 | ✓ | ✓ |
| Timber legality assurance systems | 4 | 2003–07 | ✓ | |
| Public procurement policies | 17 | 2001–10 | | ✓ |
| Certification schemes | 4 | 1989–2001 | | ✓ |
| Green Building Initiative | 3 | 1996–98 | | ✓ |
| Trade initiatives to promote legal and sustainable timber | 5 | 1991–2008 | ✓ | ✓ |
| NGOs for advocacy | 2 | 1995–2000 | ✓ | ✓ |
| NGOs for legal timber production | 5 | 1990–2000 | ✓ | ✓ |
| Independent forest monitoring | 4 | 1984–2003 | ✓ | |
| Regional and country support facilities | 3 | 2002–08 | ✓ | ✓ |
| Knowledge development and capacity building | 14 | 1985–2010 | ✓ | ✓ |
| **Total** | **122*** | | | |

FLEG – Forest Law, Enforcement and Governance; FLEGT – Forest Law Enforcement, Governance and Trade; NGO – non-governmental organization

*This excludes five general fact sheets

## 2.7 EXTENT OF FOREST CERTIFICATION: STATUS REVIEW

Certified wood has yet to achieve a critical mass. A sustainable change would require it to capture at least 13% of the share in the marketplace. In fact, by some measures, the expansion of certification in recent years has been driven primarily by the paper industry, and the significant impact is, thereby, limited to this sector of the marketplace. Certification still needs to establish itself more broadly in the solid-wood products marketplace. In the final analysis, it appears that the harmonization of chain of custody (CoC) requirements for multiple certification systems could go a long way towards helping the overall market grow (Fernholz, Howe, Bratkovich, *et al.* 2010). The temporal status of certification and its extent are discussed below.

### 2.7.1 Status in 2003

There were a variety of third-party certification organizations in the world, and over 125 Mha (>3% of the world's forests) were certified by one or more organizations (Rametsteiner and Simula 2001). However, the majority of these certified forests were in North America and Europe.

### 2.7.2 Status in 2005

There was also considerable variation in the types of timber and the timber sectors that were most heavily certified. Global statistics for certification were not available; yet a report commissioned by the UK Timber Trade Federation showed that 56% of timber products imported into the UK in 2005 were certified. The proportion of certified wood across different products varied considerably—for instance, 98% of imported medium-density fibreboard, 58% of imported sawn softwood, and only 11% of sawn hardwood imports were certified. These sort of differences reflected variations in the regions from where the products originated from and the relative engagement of different sectors with certification schemes.[1]

### 2.7.3 Status in 2006

Globally, about 271 Mha of forest had been certified as of January 2006. Combined together, FSC (25.14%) and PEFC (68.69%) accounted for 93.8% of all certified forest area. The vast majority of certified forests were in the temperate and boreal regions of North America and Europe, which together account for 91.8% of the total spread. Developing countries accounted for just 13% of certified forests, while tropical forests—the original focus of certification—had just 4.7% of forest area as certified. Tropical developing countries with the largest areas of certified forests included Brazil, Bolivia, Mexico, and Guatemala. Also the vast majority of certified forest areas (both tropical and non-tropical) were industrial forests (Durst, McKenzie, Brown, *et al.* 2006). Table 2.3 summarizes the extent of certification of different schemes in the world and within the USA.

---

[<?>] Details available at <www.forestry.about.com/od/certification/Forest_Certification.htm>

**Table 2.3** Forest certification systems around the world and within the USA in 2008

| *Certification system* | *Forest area certified in the world (Mha)* | *Forest area certified in the USA (Mha)* |
|---|---|---|
| Sustainable Forestry Initiative | 59.9 | 21.7 |
| Forest Stewardship Council | 92.9 | 9.0 |
| American Tree Farm System | 9.7 | 9.7 |
| Canadian Standards Association | 79.3 | 0.0 |
| Programme for the Endorsement of Forest Certification | 204.0 | 0.0 |

**Source** Bettinger, Boston, Siry, *et al.* (2009)

Within the USA, approximately 16 Mha of certified forests were located in the southern states. It is interesting that the Southern USA contained about 29% of the nation's forests and about 46% of certified forests. This is intriguing since federally managed forests are more commonly found in the western portion of the nation, and these are not certified. The forest industry owned about 9 Mha of pine plantations in 2005, and almost all this land would have been certified. However, much of the forest industry land has changed ownership in the past few years, mainly through sales from vertically integrated forest product companies to large timber investment management organizations (TIMOs) or real estate investment trusts (REITs). What effects this change in ownership might have had on the amount of certified forest area is interesting to analyse (Bettinger, Boston, Siry, *et al.* 2009).

## 2.7.4 Status in 2009

Approximately, 8% of the global forest area was certified under a variety of schemes (FAO 2009b). A recent estimate suggests that approximately one-quarter of global industrial roundwood came from certified forests (FAO 2009b). Most of these advances had occurred outside the tropics—less than 2% of African, Asian, and tropical American forests are certified. Most certified forests (82%) were large and managed by the private sector (ITTO 2008). An overview of the global supply of roundwood from certified sources is given in Tables 2.4 and 2.5.

**Table 2.4** An overview of the global supply of roundwood from certified sources in 2007/08

| *Region* | *Total forest area (Mha)* | *Total forest area (Mha)* | | | *Total certified forest area (%)* | | | *Industrial roundwood produced from certified forest (million $m^3$)* | | | *Industrial roundwood produced from certified forest from global roundwood production (%)* | | |
|---|---|---|---|---|---|---|---|---|---|---|---|---|---|
| | | *2007* | *2008* | *2009* | *2007* | *2008* | *2009* | *2007* | *2008* | *2009* | *2007* | *2008* | *2009* |
| North America | 470.6 | 164.2 | 181.7 | 180.3 | 34.9 | 38.6 | 38.3 | 210.1 | 232.5 | 230.7 | 13.2 | 14.6 | 14.5 |
| Western Europe | 155.5 | 80.8 | 84.2 | 82.2 | 52.0 | 54.1 | 52.8 | 166.4 | 173.4 | 169.2 | 10.5 | 10.9 | 10.7 |
| CIS | 907.4 | 20.6 | 24.6 | 25.2 | 2.3 | 2.7 | 2.8 | 3.6 | 4.3 | 4.4 | 0.2 | 0.3 | 0.3 |
| Oceania | 197.6 | 9.9 | 9.4 | 10.3 | 5.0 | 4.8 | 5.2 | 2.5 | 2.4 | 2.6 | 0.2 | 0.1 | 0.2 |
| Africa | 649.9 | 2.6 | 3.0 | 5.6 | 0.4 | 0.5 | 0.9 | 0.3 | 0.3 | 0.6 | 0.0 | 0.0 | 0.0 |
| Latin America | 964.4 | 12.1 | 15.0 | 14.6 | 1.3 | 1.6 | 1.5 | 2.1 | 2.6 | 2.5 | 0.1 | 0.2 | 0.2 |
| Asia | 524.1 | 1.6 | 2.0 | 3.0 | 0.3 | 0.4 | 0.6 | 0.7 | 0.8 | 1.3 | 0.0 | 0.1 | 0.1 |
| **World total** | **3869.5** | **291.8** | **319.9** | **321.2** | **7.5** | **8.3** | **8.3** | **385.7** | **416.4** | **411.3** | **24.3** | **26.2** | **25.9** |

CIS – Commonwealth of Independent States

**Note** The reference for forest area (excluding "other wooded land") and estimations for industrial roundwood production from certified forests are based on *State of the World's Forests 2009* (FAO 2009b). Concerning roundwood production, the subregions' annual roundwood production from "forests available for wood supply" is multiplied by the percentage of the subregions' certified forest area (it is assumed that the removal of industrial roundwood from each hectare of certified forests is the same as the average for all forest available for wood supply). However, not all certified roundwood is sold with a label. The 2008 and 2009 figures are from May to May.

**Sources** Individual certification systems; Forest Certification Watch; Certification Canada; FAO (2009b); Oliver 2010

**Table 2.5** Global supply of industrial roundwood from certified resources (2008–10)

| *Region* | *Total forest area (Mha)* | *Certified forest area (Mha)* | | | *Certified forest area (%)* | | | *Volume of industrial roundwood from certified forest (million $m^3$)* | | | *Proportion of global roundwood production from certified forests (%)* | | |
|---|---|---|---|---|---|---|---|---|---|---|---|---|---|
| | | *2008* | *2009* | *2010* | *2008* | *2009* | *2010* | *2008* | *2009* | *2010* | *2008* | *2009* | *2010* |
| North America | 613.2 | 181.7 | 180.3 | 199.8 | 38.6 | 29.4 | 32.6 | 232.5 | 175.6 | 194.6 | 1.46 | 9.8 | 10.9 |
| Western Europe | 166.2 | 84.2 | 82.2 | 85.0 | 54.1 | 46.5 | 51.2 | 173.4 | 238.1 | 261.7 | 10.9 | 13.3 | 14.6 |
| CIS | 835.3 | 24.6 | 25.2 | 29.9 | 2.7 | 3.0 | 3.6 | 4.3 | 4.9 | 5.8 | 0.3 | 0.3 | 0.3 |
| Oceania | 206.3 | 9.4 | 10.3 | 11.6 | 4.8 | 5.0 | 5.6 | 2.4 | 2.5 | 2.8 | 0.1 | 0.1 | 0.2 |
| Africa | 635.4 | 3.0 | 5.6 | 7.3 | 0.5 | 0.9 | 1.2 | 0.3 | 0.6 | 0.8 | 0 | 0.0 | 0.1 |
| Latin America | 924.2 | 15.0 | 14.6 | 14.4 | 1.6 | 2.1 | 1.6 | 2.6 | 3.6 | 2.7 | 0.2 | 0.2 | 0.1 |
| Asia | 571.4 | 2.0 | 3.0 | 8.6 | 0.4 | 1.4 | 1.5 | 0.8 | 3.1 | 3.4 | 0.1 | 0.2 | 0.2 |
| **World total** | **3952.0** | **319.9** | **321.2** | **156.7** | **8.3** | **8.2** | **9.0** | **416.4** | **428.4** | **471.8** | **26.2** | **24.0** | **26.4** |

CIS – Commonwealth of Independent States

**Note** Information valid as on May 2010.

**Sources** Individual certification systems; Forest Certification Watch; Certification Canada; FAO (2009b); Oliver 2010

### 2.7.5 Status in 2010

By May 2010, the area of certified forests endorsed globally by one or other of the international certification frameworks—FSC or PEFC—amounted to 355 Mha. This represents an increase of approximately 28 Mha (8%) since January 2009, which is in line with recent trends. After an increase in global certified forest area of 125 Mha in 2004/05 in response to industry-wide commitments in North America, the pace of growth in certification has slowed to an average of 23 Mha/year since 2006 (Oliver 2010).

In mid-2010, FSC approved standards existed in 13 countries, with interim standards (Generic Forest Stewardship Standards) in 66 further countries. FSC recently reached a milestone when the number of forest management certificates issued worldwide exceeded 1000 for the first time. While the FSC has now penetrated most areas of the world, gaps remain in parts of tropical Africa and South East Asia. The largest areas of FSC-certified forests are in North America, Sweden, and Russia (Oliver 2010).

### 2.7.6 Certified Forests

The area of certified natural forest production, that is, public forest estate (PFE) increased in each region between 2005 and 2010. In Asian, Asia-Pacific, and Latin American regions together, the certified forest area grew from 10.5 Mha to 17.0 Mha, an increase of 63% (1.3 Mha/year). In percentage terms, the biggest growth was in Africa, where the certified forest area more than tripled, from 1.48 Mha to 4.63 Mha (ITTO 2011).

The general upward trend in the area of certified forest declines in some countries. In Bolivia, for example, there was a decline of about 500 000 ha between the two surveys, and in Mexico there was a drop of about 150 000 ha.

### 2.7.7 Area under Sustainable Forest Management, Estimates of Trend

The area of production of PFE considered under sustainable forest management increased between the 2005 and 2010 surveys from 25.2 Mha to 30.6 Mha, an increase of about 20% (1.1 Mha/year). This was despite a significant decline in the area under sustainable forest management in Papua New Guinea (where the estimate made in the 2005 survey was likely a significant overestimate) and lesser decreases in several other countries, such as Central African Republic, Côte d'Ivoire, and Ghana. The area was steady in Asia and the Pacific and increased in Africa and Latin America and the Caribbean. Even though the estimated total area of natural forest production of PFE is somewhat larger than the area estimated in 2005, the area under sustainable forest management as a percentage of the natural forest production of PFE increased slightly, from 7.1% in 2005 to 7.6% in 2010 (ITTO 2011).

# 3
CHAPTER

# Schemes for Certification

A wide range of certification schemes are available for forest growers/ owners, and these involve process or performance standards and have widely different objectives. It is essential for forest owners and organizations to identify their objectives in seeking forest certification and select an appropriate certification scheme to meet these objectives (Handford and Nussbaum 2001).

## 3.1 TYPES OF CERTIFICATION SCHEMES

Certification schemes can be broadly categorized into two major groups: (1) performance based and (2) process based (Layton, Guynn, and Guynn 2002; Perera and Vlosky 2006).

Performance-based standards define specific performance levels for various aspects of forest management. It is the framework used by organizations like the Forest Stewardship Council (FSC).

Process-based schemes provide a systematic approach to developing, implementing, monitoring, and evaluating environmental policies, but they do not stipulate performance standards and use the framework used by the Sustainable Forestry Initiative (SFI) (Layton, Guynn, and Guynn 2002; Perera and Vlosky 2006).

The two certification processes may differ in the ways that they approach conservation and management of habitat for various forestry purposes. In both approaches, the matrix used to define compliance or non-compliance may be areas of concern for forest owners contemplating certification (Perera and Vlosky 2006). Box 3.1 elaborates on forest certification standards.

**Box 3.1** Forest certification standards

In the World Trade Organization context, forest certification standards fall into the category of process and production method standards, which specify how natural resources are managed and raw materials are harvested. Both performance and management systems are quality aspects covered by sustainable forest management standards. They define those measures or outcomes of forestry interventions that are acceptable and the kind of management system that should be in place before a certificate can be issued (Atyi and Simula 2002).

However, any credible third-party certification requires three basic ingredients (Handford and Nussbaum 2001): (1) standard against which organizations are assessed; (2) defined certification process; and (3) accreditation—access to assess certification bodies.

Although there are many schemes of forest certification prevalent over the globe, two principal schemes of forest certification—FSC and Programme for the Endorsement of Forest Certification (PEFC)—are very important in terms of coverage.

### 3.1.1 Possible Ways to Ensure Certification

Countries around the world accept the concept of certification because of the reasons discussed in the subsequent sections.

#### 3.1.1.1 *Market driven*

Forest certification emerged in the mid-1990s as a market-based incentive to improve forest management (Cashore and McDermott 2004; Vogt, Larson, Gordon, *et al.* 2000). The countries involved in timber trade accepted it because it had become the demand of the market. The green consumer now exercises significant purchasing decisions in favour of "green woods". Therefore, self-interest of the cooperate sector in the face of declining resources and increasing client power have created conditions for market-based certification programmes to succeed (Upton and Bass 1995). The market demand for certified products is gradually increasing in the international market as discussed in detail in Chapters 2 and 4.

Market-based, voluntary approaches have the following advantages.

- They can yield cost savings over regulatory mechanisms by allowing forest managers to determine the most appropriate and cost-effective ways of meeting given external standards.

- They offer an ongoing incentive to continuously improve and, in doing so, encourage the development of new techniques and technologies.
- They are flexible, allowing forest managers to work within overall constraints and do not require the drafting of new legislations and regulations by the government.
- They are relatively efficient in terms of resource pricing and thus can promote conservation and long-term sustainable resource use.

However, there are some operational dilemmas, which are as follows.

- There is much more experience of using regulation than economic instruments to achieve environmental and social goals. Economic instruments have been much talked about but less implemented.
- Economic instruments are unlikely to work on their own in low-income or non-market situations where people are too poor to pay and markets are undeveloped, facing many uncertainties about supply and demand and macroeconomic instability.
- In the past, the market can hardly be said to have been the friend of the forest. Rather, the commoditization of just one or two forest goods—wood and land—has led to widespread forest loss and degradation. According to some observers, this has been the main cause of forest loss and degradation (Upton and Bass 1995).

Hence, regulatory and institutional provisions are required as a supplement to market-based initiatives such as certification to correct these market failures (Upton and Bass 1995).

#### 3.1.1.2 *Social responsibility*

Forest certification was initially adopted by most European countries as an indication of society's concern to curb the increasing rate of deforestation. Because of high societal awareness and concern towards the increased rate of deforestation in tropical forests, the concept of forest certification emerged in the 1980s as a possible tool to control deforestation and enhance the social and environmental importance of tropical forests in particular.

As a voluntary mechanism, forest certification is considered an addition to the command and control method of governance, the latter normally being enacted by governments through laws, regulations, and the best available technology prescriptions. However, social pressures often pose more requirements than do governments. Some forest managing entities have accepted the public pressure to adopt forest certification, while others have exercised caution in embracing the process (Leslie 2004).

#### 3.1.1.3 Regulatory mechanism

Certification plays an important role in the thrust to save forests through the public procurement policy of national governments, allowing timber buyers in the public sector to purchase timber that meets government standards. Besides this, the Lacey Act and the European Union Forest Law Enforcement, Governance and Trade (EU FLEGT) have come up with further stringent measures to prevent trade in illegal timber. This will be a major driver of assessing illegal/ unsustainable deforestation and the cause of severe social, economic, and environmental damage. The details of the Act are given in Annexes 2 and 3. Other examples of regulatory mechanisms are as follows.

***Public procurement policies of timber products in other countries***

The Global Broadband and Innovations Programme, a green public procurement policy for forest products, is a relatively new instrument and its implementation is still at an early phase in most countries. By May 2010, at least 12 national governments had operational green procurement policies, including specific criteria for forest products. These include eight countries in Europe (Belgium, Denmark, France, Germany, the Netherlands, Norway, Switzerland, and the UK).

There have been significant developments at local levels also. For example, the Olympic Delivery Authority has specified that timbers certified by the PEFC and FSC will be used for all constructions, including buildings, transport, and energy infrastructure for the 2012 London Olympics. In the Russian Federation, the Ministry of Natural Resources recently endorsed "green standards" for the 2014 Winter Olympics in Sochi, which specify that FSC-certified timber will be used.

## 3.2 VARIOUS SCHEMES FOR FOREST CERTIFICATION

Forest certification schemes have been categorized into internationally recognized schemes and nationally recognized schemes. These schemes are summarized in Table 3.1.

The internationally recognized schemes, such as FSC, SFI, Canadian Standards Association (CSA), and PEFC, have wider acceptability in the international market in comparison to national schemes that are designed to suit the market requirements of a specific country or region. In comparison to international schemes, national schemes have a limited market but are less expensive and more easily implementable; for example, the Keurhout Foundation in Holland and the Malaysian National Timber Certification Council in Malaysia have been initiated as country-level certifying bodies (Rametsteiner and Simula 2001).

Details of some of the important internationally and nationally recognized schemes are given in the subsequent section.

### 3.2.1 International Organization for Standardization

In 1993, the ISO 14000 Series on Environmental Management Systems was announced by the International Organization for Standardization

**Table 3.1** Some national and international certification schemes

| *Internationally recognized scheme* | *Nationally recognized scheme* | |
|---|---|---|
| | *Name of scheme* | *Country* |
| Forest Stewardship Council (FSC) | Canadian Standards Association (CSA) | Canada |
| Programme for the Endorsement of Forest Certification (PEFC) | American Tree Farm System (ATFS) | America |
| International Organization for Standardization (ISO) 1400 | Australian Forestry Standard (AFS) | Australia |
| Sustainable Forestry Initiative (SFI) | Brazilian National Forest Certification Program (Cerflor) | Brazil |
| | Chilean Forest Certification System (CERTFOR) | Chile |
| | Indonesian Ecolabeling Institute (LEI) | Indonesia |
| | Malaysian Timber Certification Council (MTCC) | Malaysia |
| | Keurhout Foundation | Netherlands |
| | China Forest Certification Scheme (CFCC) | China |

(ISO) and received strong support from the paper and pulp industries. This initiative was a potential counterbalance against the FSC. The differences in the dominant approaches being taken by both the groups resulted in protocols being classified as a performance standards approach or an environment management systems approach (Vogt, Larson, Gordon, *et al.* 2000).

## 3.2.2 Forest Stewardship Council

### 3.2.2.1 *Background*

In response to varying levels of certification credibility, the FSC was established in 1993 (Perera and Vlosky 2006). It originated in Europe and is active in 57 countries around the world, including the USA.[1] The FSC secretariat is in Oaxaca, Mexico. It was initiated and established by the World Wide Fund (WWF). It is an international, independent, non-profit, non-governmental organization (NGO). It has nearly 300 members from over 40 countries.

The FSC was organized to evaluate, accredit, and monitor certifiers and to strengthen the process through training, education, and development of national initiatives. It does not certify forest products itself; instead, it gives consumers the assurance that certifying organizations have the highest level of credibility and integrity.

The FSC is an international organization that evaluates, accredits, and monitors independent forest product certifiers. By accrediting certifiers, it assures consumers that certification labels are consistent and reliable and adhere to an international set of forest stewardship principles. It is funded by accreditation fees, membership dues, royalties for using its logo, and charitable donations. The activities of FSC are orientated towards the following.[2]

- Improving forest management.
- Incorporating full costs of management and production into the price of forest products.
- Promoting the highest and best use of forest resources.
- Reducing damage and waste.
- Avoiding overconsumption and overharvesting.

[1] Details available <www.fsc.org>

[2] Details available at <www.fsc.org>

The general objectives of FSC ensure the following.[3]

- Environmentally appropriate forest management; enabling the harvest of timber and non-timber products while maintaining the forest's biodiversity, productivity, and ecological processes.
- Socially beneficial forest management to ensure long-term benefits to society and provide strong incentives to local people to sustain forest resources.
- Operational economic viability; ensuring forest operations are sufficiently profitable to ensure stability of operations and a genuine commitment to the principles of good forest management.

#### 3.2.2.2 *Operational structure and organization*

Operational implementation of FSC has been devolved to country-level operations called national initiatives, wherever possible. Decentralization enhances the role for national initiatives, which take on responsibility for the development of standards, national fund-raising, national marketing, and national membership services and where possible, play an integral role in the accreditation process. Their role is formalized through agreements between FSC AC in Mexico and each national initiative.

#### 3.2.2.3 *Certification principles, criteria, and procedures*

The FSC does not prohibit or endorse any silvicultural practice. Instead, certification under FSC requires that the overall health and integrity of the forest ecosystem is conserved and maintained. This involves ensuring regeneration of the forest, conservation of its biological diversity, and maintenance of its natural cycles (such as nutrient recycling) to protect the long-term health and productivity of the forest. These objectives can be reached through a variety of silvicultural practices (Fernholz, Howe, Bratkovich, *et al.* 2010).

The principles and criteria apply to all tropical, temperate, and boreal forests and are, therefore, quite general. There are specific guidelines for commercial plantation forestry. Accredited certifiers use site-specific forest management standards for the evaluation of individual forests. Site-specific forest management standards are developed through a multi-stakeholder consultative process consistent

---

[3] Details available at <www.fsc.org>

with all applicable laws and FSC principles (Fernolz, Howe, Bratkovich, *et al.* 2010).

A recent study conducted by Dovetail Partners gives an indication of the trend in changes that different certification schemes have undergone since 2004. As per the report, the FSC has undergone the following changes in a period of six years (Fernolz, Howe, Bratkovich, *et al.* 2010).

### 3.2.2.4 *Trend of changes in FSC over the past few years*

Changes within the FSC programme in the past six years include the following.

- Development of a volume-credit-based chain of custody (CoC) control system.
- Development of a Controlled Wood Standard for evaluating the non-FSC content of a mixed source product or material.
- Establishment of an independent auditor accreditation services through the Accreditation Services International (ASI).
- Establishment of a requirement to issue non-conformances at the indicator level during forest management assessments.
- Revision of the nine regional standards in the USA to create a single national standard (currently in final draft form and awaiting approval from the FSC International Centre).
- Development of Family Forest Indicators within the US National Standard (currently in draft form, having completed the first round of public comment).
- Revision of the Group Certification Standard to allow greater flexibility in demonstrating compliance with some of the indicators at the group level, rather than at the individual group member level.

In general, changes to the FSC programme have been made to accommodate unique challenges and needs while strengthening the accountability and integrity of the standards. The FSC programme is still challenged by a perception that compliance costs for small businesses and landholders are too high. Additionally, some business interests view the standards and requirements for the FSC programme as changing too frequently.

Since 2004, global FSC certification has grown from 125 million to 304 million third-party-certified acres and a total of 987 forest management certificates.

Figure 3.1 illustrates the distribution of FSC-certified forestlands globally. In terms of CoC certification, the FSC has grown from 4000 CoC certificates in 2004 to 16,235 today. In 2004, the FSC was operating in 62 countries, whereas the number has increased to 80 today.[4]

The FSC-certified forest area stood at 129 million hectares in May 2010, up from 107 million hectares in January 2009.

Much of the increase in FSC-certified area during the period took place in North America (more than 10.7 million hectares) and the Russian Federation (more than 4 million hectares) (Oliver 2010). FSC-certified forests are distributed in more than 80 countries around the world—Africa, Asia and Oceania, Europe, North America, and Central and South America. They cover all forest biomes—boreal, temperate and tropical, and subtropical—and are managed under all types of ownership—public, private, concessions, and community. With some 15,000 certified enterprises along the supply chain committed to FSC, the continued market growth is sending a clear message to forest managers that responsible forest management is the smart way forward (FSC 2009b). Box 3.2 enumerates the types of eco-labelling.

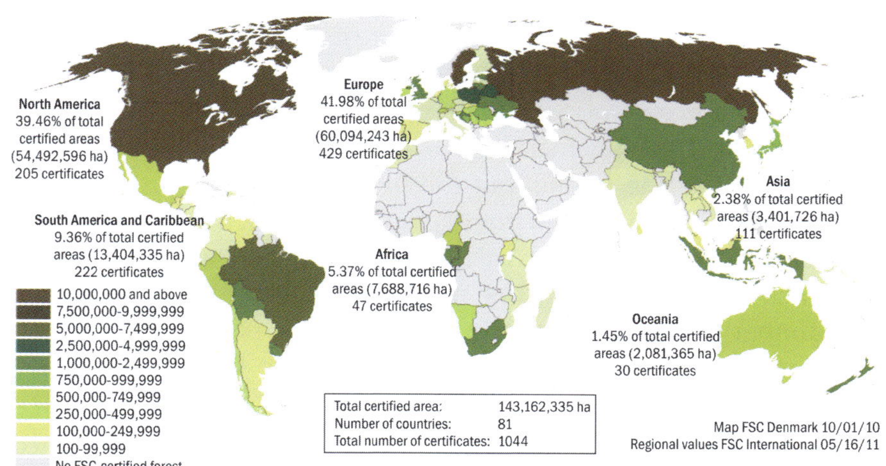


**Figure 3.1** Global FSC-certified forest area by region
**Source** Global FSC certificates: types and distribution. 2010

[4] Details available at <www.fsc.org/facts-figures.html>

**Box 3.2** Eco-labelling

A company may want to inform consumers about the environmental claims of a specific product or service through the use of eco-labels. Eco-labelling is a voluntary certification and verification process.

The International Organization for Standardization (ISO) classifies three broad types of eco-labels.

- *Type I* A voluntary, multiple-criteria-based third-party programme that authorizes the use of environmental labels on products indicating the overall preference of a product within a particular category based on life cycle considerations. Examples include the EU Flower and the Environmental Choice Programme of Canada.
- *Type II* A programme involving self-declared environmental claims by parties likely to benefit from such claims. These programmes often involve single attributes. An example is the Paper Profile.
- *Type III* A programme involving a declaration that provides quantified environmental life cycle product information provided by the supplier, based on independent verification and systematic data presented as a set of categories of a parameter.

There are many eco-labels in the world. In addition to the Forest Stewardship Council and the Programme for the Endorsement of Forest Certification, other important eco-labels for wood and paper-based products are summarized in Annex 4.

### 3.2.3 Sustainable Forestry Initiative

In 1994, members of the American Forest and Paper Association, primarily composed of forest products companies, agreed to operate under a set of forestry principles that were suggestive or indicative of sustainable forest management (SFM). These principles were designed to demonstrate to the public that the industry could meet the needs of the present generation without compromising on the needs of future generations. The SFI, which arose from this effort, called for land ethics that integrated active management with the conservation of non-timber forest products and aesthetics. The objectives of the SFI programme have been to encourage and promote the efficient use of forest resources. The objective also called for the protection of certain wildlife habitats and areas that contribute to forest biodiversity for the protection of water quality and for managing the potential visual impact related to harvesting. The SFI uses a hierarchy of five principles, 13 objectives, and 34 performance measures within those objectives and 103 indicators (Bettinger, Boston, Siry, *et al.* 2009). To ensure long-term sustainability of forest management, organizations engaged in sustainable forestry initiatives must also develop a management plan and associated analysis that involves the following.

- Assessment of the growth of forest resources.
- Development of an inventory system to track the growth of forest resources.
- Development of an inventory system to track growth and harvest.
- Development of a land classification system.
- Development of soils inventory.
- Assessment of sustainable harvest levels.
- Review of non-timber forest products management issues.

A recent study conducted by Dovetail Partners gives an indication of the trend in changes that different certification schemes have undergone since 2004. As per the report, sustainable forestry initiatives have undergone the following changes in a period of six years (Fernholz, Howe, Bratkovich, *et al.* 2010).

- Achievement of international endorsement from PEFC schemes; international recognition of the SFI programme standard and label; transition to a fully independent 501(c) 3 structure; creation of a diversified decision-making structure representing economic, ecological, and social interests; and establishment of a CoC standard for the tracking and labelling of SFI raw materials and products.
- Transition to a full third-party auditor-based programme.
- Accreditation through the American National Standards Institute (ANSI).
- Development of public reporting, including publicly available assessment reports.
- Incorporation of social considerations within the standard, including indigenous people's rights and stakeholder consultation.
- Expansion to include Canada within the scope of SFI standard.
- Completion of a five-year standard revision process, including opportunity for public comment (Fernholz, Howe, Bratkovich, *et al.* 2010).

Overall, changes in the SFI programme have transformed a national, industry-led effort into an internationally recognized and independently managed programme. The impact of the SFI programme has grown from approximately 90 million acres of third-party-certified forestland in 2004 to 181 million acres today, and from having issued no CoC certificate in 2004 to 807 CoC certificates issued by the end of 2009. Despite significant changes and growth, the SFI programme is still not fully recognized in most green building programmes, including the Leadership in Energy and Environmental Design ratings systems managed by the United States Green Building Council (USGBC).

### 3.2.4 Programme for Endorsement of Forest Certification

European forest industry organizations have formally launched a certification scheme for sustainably produced timber, which will compete with the global WWF-backed FSC. At a meeting in Paris, representatives of 12 national bodies participating in the USGBC signed the new body's statutes. The development of PEFC originated with small forest owners in Finland and some other European countries who opposed the FSC certification scheme. Organizations from 17 European countries are now participating in the PEFC, and organizers say that expressions of interest have been received from forest industry organizations in Australia, Canada, USA, and Brazil.[5]

According to PEFC, it now has the support of associations representing about 12 million woodland owners in Europe, who manage 100 million hectares of woodlands and cut 280 million $m^3$ of timber annually.

The scheme provides a framework for mutual recognition of national forest certification schemes. PEFC explicitly allows for participation by all stakeholders, including environmental groups, retailers, as well as forest owners and industries. Eleven countries have now established national PEFC governing bodies: Austria, Belgium, Czech Republic, France, Finland, Ireland, Norway, Portugal, Spain, Sweden, and Switzerland.

---

[5] Details available at <www.pefc.org/standards/national-standards/certification-accreditation-requirements>

#### 3.2.4.1 *Criteria and process*

PEFC has defined common elements and requirements that have to be met by certification schemes wishing to take part in and use the trade mark of PEFC. These minimum requirements will help to promote SFM and assure consumers that products with PEFC label come from sustainably managed forests.[6]

The certification criteria cover the whole range of forest functions, including all economic, ecological, and social functions. The certification criteria to be used in PEFC are based on the six pan-European criteria for SFM as a common framework. National criteria will respect the relevant legal requirements, national policies, and programmes.[7]

The Pan-European Process has identified a set of 27 quantitative and descriptive indicators for national monitoring and reporting. Member countries are free to elaborate these indicators at the national level. Since these guidelines and SFM indicators are not relevant under all local conditions, they serve as a framework, which will be interpreted as appropriate in national or regional situations.[8]

The certification criteria are expected to cover all relevant aspects of SFM and administrative conditions and systems. To ensure a sufficient degree of equivalency and comparability between admissible certification schemes, minimum requirements have been defined for the key aspects of forest certification.[9]

Specific certification criteria can be established and adopted at national, regional or other sub-national levels by forest units. All relevant interested parties need to be invited to participate in this process in an open and transparent way. The development of certification criteria has to be independent from the certification process. The criteria will be periodically reviewed and updated to

---

[6] Details available at <www.pefc.org/images/stories/documents/external/Full_procurement_guide_v1.1_final.pdf>

[7] Details available at <www.pefc.org/standards/national-standards/certification-accreditation-requirements>

[8] Details available at <www.pefc.org/standards/national-standards/certification-accreditation-requirements>

[9] Details available at <www.pefc.org/images/stories/documents/external/Full_procurement_guide_v1.1_final.pdf>

reflect new scientific knowledge and the results must be made public. While consensus is the objective, it is not a precondition to decide on criteria.

The level of application for certification under PEFC can be at a regional, group or individual level so long as all members of a region or group comply with the requirements of the relevant certification scheme.

The standards and requirements for the auditing and certification procedures of PEFC are mainly based on EN 45011 (general requirements for bodies operating product certification systems), EN 45012 (general requirements for bodies operating assessment and certification/registration of quality systems), and EN 30011-2 (qualification criteria for environmental auditors).[10] A certification process consists of five steps.

1. Application for certification by the forest owner
2. Assessment process
3. Reporting
4. Decision on certification
5. Re-auditing

Certification bodies are foreseen to be accredited by national accreditation bodies or involve accredited certification organizations. PEFC-approved certification bodies are independent third parties that assess and certify forest management and/or the performance of the forest in the unit to be certified with respect to the certification criteria.[11]

PEFC has grown significantly in the last six years. When Dovetail first reported on PEFC, there were 16 endorsed programmes and 28 national forest certification programmes participating as PEFC members. In 2004, its endorsed programmes represented about 130 million acres of certified forestland (Fernholz, Howe, Bratkovich, *et al.* 2010). Today there are 28 endorsed national programs and 34 PEFC members. It now consists of 544 million acres of certified forestland within the endorsed programmes, and nearly 6200 PEFC-CoC certificates have been issued.[12] Additionally, as of February

---

[10] Details available at <www.pefc.org/images/stories/documents/external/Full_procurement_guide_v1.1_final.pdf>

[11] Details available at <www.pefc.org/standards/national-standards/certification-accreditation-requirements>

[12] Details available at <www.pefc.org/index.php/about-pefc/who-we-are/facts-a-figures>

2010, PEFC has announced a policy to allow stakeholders to become members.

Regional certification is an approach developed by PEFC to address non-industrial forestland in Europe. It allows a landowner organization to apply for certification on behalf of its members. This approach has been used in Finland, where 60% of forests are privately owned and a network of forestry associations is available to support certification and management activities (Oliver 2010).

Regional certification has also been applied by PEFC-managed programmes in Austria, Belgium, Czech Republic, France, Germany, Italy, Latvia, Luxembourg, Slovakia, and Spain. There are several developments expected in non-industrial forest certification. The CSA has completed the development of a smallholder standard and is expected to seek PEFC endorsement. The FSC programme has moved forward with an initiative to adapt standards to address small- and low-intensity-managed forests (Oliver 2010).

## 3.3 SOME NATIONAL CERTIFICATION PROGRAMMES

### 3.3.1 Canadian Standards Association

CSA developed a performance- and system-based certification process in 1996. The process was developed as a voluntary SFM standard for lands within Canada (Bettinger, Boston, Siry, *et al.* 2009).

While CSA is national in scope, PEFC is an international umbrella organization for the assessment of national certification schemes. Again within PEFC, the criteria for national schemes are developed by their stakeholders; so they vary according to the type of forests and local sociopolitical factors (Bettineger, Boston, Siry, *et al.* 2009).

As per the study conducted by Dovetail Partners, the CSA programme was the first North American programme to become PEFC endorsed and continues to serve the interests of Canadian land managers and organizations. CSA has recently completed the development of a smallholder standard (CSA Z804) to increase certification opportunities for small properties in Canada. It is anticipated that CSA will pursue PEFC endorsement for the new smallholder standard.

At the level of individual countries, the number of certification schemes under development is increasing rapidly. Bass and Simula

(1999) provided details of 12 active schemes and reported that development work was under way in at least 12 other countries. National schemes can be divided into three main groups.

1. Schemes aligned from the outset with either FSC or PEFC.
2. Schemes that develop independently but aim for compatibility with FSC and/or PEFC.
3. Schemes without any links to an umbrella scheme.

Some bespoke national schemes are assisting others; for example, Indonesia's Lembaga Ekolabel Indonesia has been assisting Vietnam and Malaysia. The majority of national schemes are market oriented, voluntary, and performance based, but some also incorporate elements of the ISO 14001 Environmental Management Systems standard. The CSA scheme was a pioneer in this respect.

### 3.3.2 Malaysian Timber Certification Council

The Malaysian Timber Certification Council (MTCC) is an independent organization established to develop and operate the Malaysian Timber Certification Scheme (MTCS) to provide independent assessments of forest management practices in Malaysia as well as to meet the demand for certified timber products. The MTCS began operation in October 2001 using a phased approach due to the greater challenges encountered in managing complex tropical forests.

#### 3.3.2.1 *Forest management certification*

Forest management certification is a process of third-party audit of forest management practices in a forest management unit (FMU) to assess compliances with the requirements of a prescribed standard, leading to the award of certificate of management under MTCC. The standard used for forest management certification is the Malaysian criteria and indicators for forest management certification, which is developed by the multi-stakeholder National Steering Committee (NSC).

#### 3.3.2.2 *Chain of custody certification*

CoC certification is a process of third-party audit of timber product manufacturers/exporters. It assesses whether timber products manufactured/exported are derived from forests certified under MTCS and leads to an award of a CoC certificate. The standard used for CoC is the Technical Document Annex 4 of PEFC.[13]

#### 3.3.2.3 *Institutional arrangements for Malaysian timber certifications*

Under the institutional arrangements for MTCS, MTCC is the national governing body, which is in overall charge of the scheme. The certification bodies receive and process the applications for certification, conduct audits, and take decisions towards awarding CoC certificate. The certification bodies are required to obtain accreditation from the national accreditation body, which is the Department of Standards, Malaysia (Figure 3.2).

During the second phase of implementation of MTCS beginning in late 2005, MTCC had been using the Malaysian criteria and indicators for forest management certification, which contains nine principles, 47 criteria, and 96 indicators. For CoC certification, since 1 July 2008, MTCC has been using the Technical Document Annex 4 of PEFC. MTCS has been endorsed by PEFC, the largest forest certification programme, representing more than 200 million hectares of certified forests worldwide, with effect from 1 May 2009.

MTCS is the first tropical timber certification scheme in the Asia–Pacific region, and the second in the world after the Gabonese Forest Certification Scheme, to be endorsed by PEFC.[14] MS ISO Certification is explained in Box 3.3.

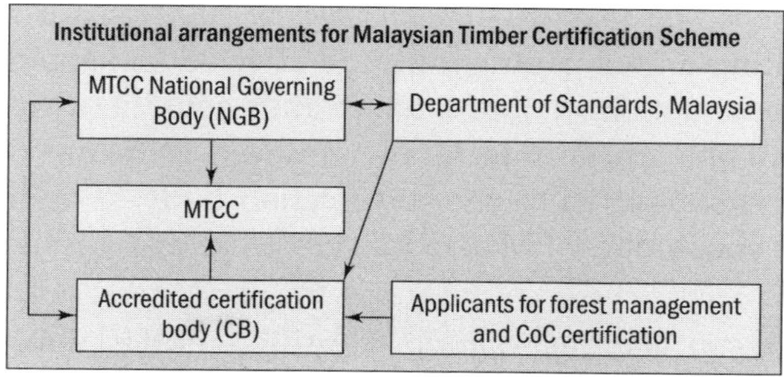


**Figure 3.2** Institutional arrangement of MTCC
**Source** <www.mtcc.com.my/mtcc_scheme_intro.asp>

[13] Details available at <www.mtcc.com.my>

[14] Details available at <www.mtcc.com.my>

**Box 3.3** MS ISO certification

The core process of certification is the production of sustainable timber from the protected reserve forest. Six major activities have been identified for certification.

- Forest boundary demarcation
- Pre-felling forest inventory
- Tree and timber tagging
- Forest harvesting
- Post-felling inventory
- Silviculture treatments

In April 2003, MS ISO 9002:1994 was awarded to the headquarters of Forestry Department Peninsular Malaysia (FDPM) and eight state forestry departments. FDPM is in the process of implementing MS ISO 9001:2008.

*Source* Discussion with Forestry Department of Peninsular Malaysia on 28/3/2011 at Kuala Lumpur

The impact of MTCC on the six core areas where the overall planning and implementation of SFM has improved as a result of forest management and certification and are summarized as follows.

1. *Documentation* Documentation review is one of the key activities of forest management and certification, which has resulted in improved compilation, maintenance, and updating of all documents related to relevant forestry manuals, legislations, guidelines, procedures, and statistics by FMUs.
2. *Consultations and stakeholders involvement* FMUs are more active in consulting and engaging with the affected stakeholders prior to the implementation of forest harvesting activities. Media reports and complaints pertaining to any certified FMU are compiled by MTCC and submitted to the certification body to be investigated. This has encouraged FMU managers to address issues of public interest.
3. *Implementation and enforcement* Certified FMUs now have well-documented guidelines and procedures for implementation in forestry activities. The certification process serves to verify that the stipulated guidelines and procedures are properly implemented on the ground and thereby helps in providing another way of overseeing the work of FMU staff and private contractors.

4. *Environmental conservation and protection* There is stricter enforcement of prescribed guidelines for road construction, forest harvesting, protection of buffer strips along waterways, and planting of open areas after harvesting. To reduce the impact of logging, FMUs are also experimenting with new environment friendly technologies such as log haulage cable system and helicopter logging. Environmental impact assessment has been conducted at a landscape level by certified FMUs in Peninsular Malaysia.
5. *Conservation and biological diversity* To monitor changes in biological diversity in harvested forests, certification process also promotes the conduct of comparative analysis between pre-felling and post-felling forest inventories. Virgin jungle reserves are also inventoried to provide baseline estimates for comparison of changes in biological diversity with comparable logged-over forest areas.
6. *Identification of gaps for research and development* Implementation of FSC has also helped in the identification of areas that required research and development inputs to further enhance the practice and implementation of SFM.

## 3.4 COMPARISON OF FOREST CERTIFICATION SCHEMES

Table 3.2 summarizes the major national, regional, and international schemes. While most cover similar ground to FSC in terms of standards and to ISO in terms of procedures, they vary in terms of participation in their development and governance and are specifically tailored to local and regional conditions. Table 3.2 gives the extent of certified forests, across major regions, as certified by major certification agencies. Annex 5 also gives a detailed comparison of features of major certification schemes.

## 3.5 DIFFERENT STANDARDS OF FOREST CERTIFICATION: STRENGTHS AND WEAKNESSES

Process standards are very useful in providing a framework for improvement. They force organizations to be systematic in their approach to understanding and addressing their impact and can create a culture of improvement through the requirements for training

**Table 3.2** Summary of main certification programmes with their salient features

| *Name of certification programme* | *Year of formation* | *Representation* | *Scope* | *Objective* | *Feature* |
|---|---|---|---|---|---|
| Sustainable Forestry Initiative (SFI) | 1994 | Members of American Forests and Paper Association (specially forest-product-based companies) operated under a set of sustaina-principles forest manag-ment Principles | USA /Canada | Encourages and promotes sustainable forestry prac-tices; efficient use of forest resources; protection of certain wildlife hibitats; protection of water quality and aesthetics | Third-party certification audit of management and practices needs to be developed; no chain of custody required<br>• Principles: 5<br>• Objectives: 13<br>• Performance measures: 34<br>• Indicators: 103 |
| Forest Stewardship Council (FSC) | 1993 | Multi-stakeholder, non-profitable organization | Worldwide | Promotes environmentally, socially, and economically viable management of forests | FSC is not a certifying organi-zation but an accrediting body that sets standards for certifica-tion organizations. It is a performance-based system. It offers three types of certification: forest management certification, chain of custody certification, and certification of controlled wood, that is, non-FSC-certified wood in processed products.<br>• Principles: 10<br>• Criteria: 56<br>• National indicators: 138 |
| American Tree Farm System (ATFS) | 1941 | Committees within ATFS | Forestland owners of the USA | Protection of environmental benefits and need for public understanding of forestry concepts | Certification standards developed by committees with ATFS; performance-based certification system; no third-party verification internal auditing system, which is |

*Contd...*

**Table 3.2** *Contd...*

| *Name of certification programme* | *Year of formation* | *Representation* | *Scope* | *Objective* | *Feature* |
|---|---|---|---|---|---|
| | | | | | required after every five years; membership limited to those lands that have passed a tree farm inspection; certified tree farms recognized by green and white signs posted along the boundaries of each farm |
| Green Tag Forestry System | | | National Forestry Association and forestry consultants serve as inspectors | | • Criteria: 10<br>• Indicators: 10 |
| Canadian Standards Association (CSA) | 1996 | | Voluntary sustainable management tool within Canada | Performance-based programme influenced by local values and public participation process | Measurement of ecosystem diversity on a broad scale; land managers need to adopt planning processes with substantial public input |
| International Organization for Standardization (ISO) 14001 | 1994 | | Worldwide | System-based process; environment management standard not specific to forest management activities but can be used | Environmental concerns identified and management plan developed accordingly; no eco-label; no chain of custody auditing |
| Programme for the Endorsement of Forest Certification (PEFC) | 1999 | | Worldwide | Performance-based certification process; includes chain of custody; eco-label for certified wood products | Chain of custody; eco-label present |

**Source** Bettinger, Boston, Siry, *et al.* (2009)

**Table 3.3** Certified forest areas (in Mha) under various certification schemes across the globe (January 2006)

| *Continent* | *FSC* | *PEFC* | *Others* | *Total* | *Per cent share* |
|---|---|---|---|---|---|
| Africa | 1.7 | 0.0 | 0.0 | 1.7 | 0.6 |
| Asia–Pacific | 2.4 | 5.2 | 4.7 | 12.2 | 4.5 |
| Europe | 35.0 | 55.9 | 0.0 | 90.9 | 33.5 |
| North America | 22.5 | 123.6 | 12.0 | 158.0 | 58.3 |
| South America | 6.5 | 1.6 | 0.0 | 8.1 | 3.0 |
| **Total** | **68.1** | **186.1** | **16.7** | **271.0** | **100.0** |
| Market share | 25.1 | 68.69 | 6.17 | 100 | — |
| Tropical | 8.0 | 0.0 | 4.7 | 12.8 | 4.7 |
| Non-tropical | 60.1 | 186.1 | 12.0 | 258.2 | 95.3 |
| Developed | 37.6 | 186.0 | 12.0 | 235.6 | 86.9 |
| Developing | 3.5 | 0.0 | 4.7 | 35.2 | 13.1 |

**Source** Durst, McKenzie, Brown, *et al.* (2006)

and monitoring (Handford and Nussbaum 2001). However, process standards do not provide any guarantee of performance as is made clear in the introduction to ISO 14001. As a result, ISO does not allow labelling of products on the basis of ISO 14001 certification. This is to avoid any confusion since people tend to equate labels with guarantees (Handford and Nussbaum 2001).

On the other hand, performance standards do guarantee a minimum level of performance and, therefore, can be linked to a product label. This is very important for many customers of certified forests as they wish to make transactions to sell labelled products. Performance guarantee is also very important in building credibility among NGOs (Handford and Nussbaum 2001).

However, there are two problems with performance standards. First, they are difficult to design to be equally applicable to all forest types, sizes, locations, and socio-economic contexts. Second, they do not include a mechanism to recognize progress made prior to the standard being met and a certificate awarded. In addition to this, for large organizations in particular, the huge number of specific criteria that must be addressed in meeting a performance standard can be both daunting and confusing (Handford and Nussbaum 2001) and not very easy. There are other trade-offs involved in this combination.

A solution to this is to combine a process standard (which provides the framework and support) with a performance standard (which provides the guarantee of a certain level of performance and thus creates a hybrid of the two approaches) (Handford and Nussbaum 2001). This too is not very easy and other trade-offs are involved in this combination.

## 3.6 COMPARING TWO MAJOR INTERNATIONAL SCHEMES

Various forest certification schemes define SFM through their respective standards. All types of forests can be sustainably managed—from primary or natural forests to intensively managed forest plantations. Annex 5 highlights how two major international schemes address the selected aspects of SFM in their approach.

## 3.7 CHOOSING A CERTIFICATION SCHEME

Each scheme uses its own label in marketing, and it would be difficult for producers, buyers, and consumers to establish which certificate should be considered reliable. Buyers cannot be expected to make an assessment of the credibility of each certificate; the task is complex and requires expertise and information that they do not usually have (Atyi and Simula 2002).

Over the past 10 years, a multitude of schemes for forest certification have been developed, making it extremely difficult for forest managers to know which SFM standards are most appropriate. There are subtle differences between the standards, but meeting different standards may require different forest management practices.

Any choice of certification scheme must take into consideration the costs of certification under different schemes. In some countries, certification under a national scheme may entail lower cost because local accreditation and certification bodies operate more cheaply than international ones. However, a national certification scheme may not provide sufficient credibility for international markets.

## 3.8 CONCLUSIONS

The following considerations need to be taken care of while deciding the type of certification scheme to be followed (Higman, Mayers,

Bass, *et al.* 2006). Forestland can be sustainably managed even without being certified by a forest certification system. Producers may not pursue forest certification if they perceive the costs of the process as outweighing the price premium offered for certified products. "Legally harvested" does not necessarily mean "sustainably produced" or "sustainably managed" because laws are sometimes insufficient to guarantee SFM or are inadequately enforced. Major certification schemes are developing methods to assess the risk of wood from non-certified sources having been produced in an unacceptable way.

# 4
CHAPTER

# Potential and Status of Certified Forest Products

## 4.1 MARKET FOR CERTIFIED FOREST PRODUCTS

The market for certified forest products is limited by the following constraints.

### 4.1.1 Limited Availability

The basic question that haunts every grower or producer of timber, or for that matter every forest manager, who intend to get his or her forest certified is about the market of their products. There are many agencies that are actually certifying timber products at present, but a limited quantity of timber being produced worldwide is certified.[1]

A report presented to the Conference on Economic, Social, and Political Issues in Certification of Forest Management, held in Malaysia during 12–26 May 1996, stated that "the quantity of timber being produced worldwide that is certified is very small. It was estimated that in 1993, approximately 1.5 million m$^3$ of timber and timber products (tropical and temperate) were certified; this was less than 0.5% of the world trade in timber".[2] These are old statistics, but even current statistics are hardly exciting.

### 4.1.2 Scattered Availability

According to an article in *The Environmental Magazine*, certified "green premium" wood has, for the most part, only scattered availability.[3] This situation is more acute in the Indian context and

[1] Details available at <www.forestry.about.com/od/certification/Forest_Certification.htm>

[2] Details available at <www.forestry.about.com/od/certification/Forest_Certification.htm>

[3] Details available at <www.forestry.about.com/od/certification/Forest_Certification.htm>

poses a serious challenge to proliferation of certified forest products in the Indian market.

## 4.2 ACCESSING NEW MARKETS

Developing something new such as a certified product is costly and time consuming; so it is essential to ensure that there is enough demand for certified products at affordable prices. Consumers purchasing certified products also demand high quality, prefer particular species, and look for reliability of supply and competitive prices. Thus customers who pay a premium price for the green credentials of their purchase are unlikely to be satisfied with poor quality. It is important to consider certification schemes that are acceptable to the new potential market. The chain of custody (CoC) certification must be achievable within the timescale for accessing the new market and should be affordable relative to the expected market value (Higman, Mayers, Bass, *et al.* 2006).

## 4.3 DEMAND FOR CERTIFIED FOREST PRODUCTS

Studies have found that interest in certification among landowners of the Pacific Northwest in the USA has increased in proportion to the increase in customers requesting certified goods. Some landowners are also concerned about the perception of their land management ability (Tikina, Kozak, and Larson 2008). The studies further suggest that in areas where land management may be highly "exposed to regulatory requirements" (especially those related to riparian management zones), landowners may have a higher interest in obtaining forest certification so as to receive recognition for the effort required to manage their land within the broader socio-economic environment. Although all forestland management organizations claim to manage land consistent with environmental rules and regulations, not all of them do so. Thus certification provides a method and means to prove their claims (Bettinger, Boston, Siry, *et al.* 2009). However, many indicators suggest that so far the final consumers of wood products have not been very active in creating a demand for certified wood products and their role in the market is not expected to bring change in the short term. In fact, globally certified wood products represent a very small fraction of total forest product sales.

The other primary force behind the drive to certify forest management may be attributed to the pressure from environmental groups. The strongest demand drivers are wood product retailers who are interested in maintaining or enlarging their public image and thus may prefer to avoid direct protests from environmental non-governmental organizations (NGOs) in their retail outlets (Hansen, Forsyth, and Juslin 2000). There is less promise of a price premium for large forested areas located in developing regions of the world because these forests traditionally have been used to meet the subsistence needs of the local people. Thus there is hardly any surplus for the export market (Bettinger, Boston, Siry, *et al.* 2009).

## 4.4 MARKET DEMAND FOR CERTIFIED FOREST PRODUCTS

### 4.4.1 Extent of CoC Certification

The market for verified wood products has been very much influenced by the extent of CoC standards (Oliver 2010). The number of CoC certificates issued worldwide increased by 41% in 2009 to reach 17 800. The Forest Stewardship Council (FSC) certification has outpaced the Programme for the Endorsement of Forest Certification (PEFC) schemes, and by May 2009, there were 12 707 FSC certificates as compared to 5108 PEFC certficates (Oliver 2010).

Aside from internationally issued CoC certificates, a limited number of regional/national certification frameworks have been issued such certificates. For example, in terms of numbers, the most significant of these is from the Sustainable Forestry Initiative (SFI) programme in North America. The number of SFI certificates issued increased considerably during the course of 2008 from 100 covering 400 locations to almost 400 covering 1000 locations. In Japan, the Sustainable Green Ecosystem Council certification programme had issued 258 CoC certificates by the end of June 2008. As a new trend, a significant proportion of some forest companies are dual certified to both SFI- and PEFC-CoC standards. While the pace of increase is impressive, data also implies that engagement in CoC certification is heavily concentrated in just a few countries. Of the 5146 new FSC- and PEFC-CoC certificates issued internationally in 2008, 47% were confined to the USA and the UK. Of those issued during the year,

70% were in only five countries—the USA, the UK, Germany, Japan, and Canada. By the end of 2008, the USA and the UK accounted for 31% of all FSC- and PEFC-CoC certificates issued internationally, with much of the rest in Germany (9%), France (7%), and Japan (6%) (Oliver 2010). This gives a picture of the distorted spread over the globe.

The available data also suggests that the number of companies with FSC- and PEFC-CoC certificates is small compared to the total number of companies engaged in the wood sector. For example, Eurostat data indicates that throughout the European Union, around 9750 FSC- and PEFC-CoC certificates were issued by the end of 2008, whereas in the subregion, there are 191 000 wood-processing enterprises, 149 377 furniture enterprises, and 19 352 pulp and paper enterprises, many of which would, in theory, be eligible for CoC certification (Oliver 2010).

In practice, CoC certification tends to be the most prevalent in the supply chains of a few large consolidated business sectors such as home-improvement retailing and parts of the paper and panels industry. It is less prevalent in more fragmented sectors, which nevertheless account for a large proportion of wood consumption, including construction and furniture. The implication is that a very large proportion of the wood supplied from certified forests never makes it to market as labelled products. This negates the purpose and value of certification as a communication tool (Oliver 2010).

## 4.5 DEVELOPMENTS IN COC STANDARDS AND PROCEDURES

The market for verified wood products is influenced not only by the extent but also by the content of CoC standards. The FSC has focused heavily in recent times on the development of procedures designed to make it easier to apply product labels in those supply chains where only a relatively small proportion of raw material supplies might be demonstrably derived from certified forests (Oliver 2010).

Application of the FSC "mixed" label has required the development of "volume credit systems" together with procedures to remove controversial wood from non-certified supply chains. These procedures have been particularly valuable in expanding the use of the FSC label in situations where wood products derive from numerous small owners, only a few of which might be certified (Oliver 2010).

### 4.5.1 Feedback From Market Participants

An ongoing research study undertaken by the Forest Industries Intelligence Ltd for the UK Timber Trade Federation with the support from the UK Department for International Development is suryeving the European market to gather information directly from market participants on trends in demand. The study includes eight European Union countries—Belgium, Denmark, France, Germany, Italy, the Netherlands, Spain, and the UK and the major findings are as follows (Oliver 2010).

- The very high level of fragmentation both in the European Union timber trade and in timber's major consuming sectors—construction and furniture—continues to present a major obstacle to CoC certification and the further development of markets for environmentally labelled wood products throughout the European Union. Other significant obstacles are generally low levels of awareness of forest certification and legality verification and very low willingness to pay among end users.
- The economic downturn is generally widening the gap between environmentally proactive operators who are now more eager than ever to exploit the opportunities emerging for timber from increasing interest in sustainable construction and those who have not felt the need to consider environmental issues and who continue to sell primarily based upon price.
- Those European Union companies that have made far-reaching commitments to shift to certified wood products often see this as part of a wider process of restructuring their overall procurement practices in favour of a limited number of key suppliers and thus be able to provide the full range of quality services, of which forest certification is only one component. For a commodity like softwood and composite panels sector, high availability of FSC- or PEFC-authorized product supply in the European Union is met with only limited market requests for labelled products. Hence, the opportunities for achieving a premium in this sector are extremely limited.

- Only in the rather restricted conditions that prevail in parts of the hardwood sector and to some extent in the speciality softwood sector (such as western red cedar cladding/siding from North America) does the issue of price premiums arise. Specific requests for certified products, particularly FSC certified, might be met by limited supply.
- The highest premiums—in the range of 20% to 50% on the price of delivery to the importers yard—are being asked for FSC-certified tropical sawn hardwood from Africa and Brazil. In the temperate hardwood sector, price premiums are being sought for FSC-certified American hardwoods in the range of 5%–10%.
- Price premiums for tropical sawnwood supplied under one or other private-sector legality verification systems (such as SGS Timber Legality and Traceability Verification [TLTV], Eurocertifor, BVQi Origin and Legality of Timber [OLB] or the Rainforest Alliance Verified Legal Origin [VLO]) are typically in the range of 3% to 15%, with most at the lower end of this range.
- Generally, there is great reluctance among end users to pay premiums for certified or verified legal wood products, a situation which places significant limits on the ability of suppliers to charge more. The highest premiums for FSC-certified tropical hardwood may only occasionally be passed on when supplying high profile public-sector contracts. As a result, there are signs that some importers and manufacturers implementing green procurement policies have switched emphasis away from FSC-certified products in favour of less expensive legally verified products when sourcing from tropical supplying countries.

## 4.6 EUROPEAN MARKETS FOR CERTIFIED FOREST PRODUCTS

The European market is undoubtedly one of the biggest and most important global markets for forest products because of its role in total consumption and trade, which underlines its size and importance. In Europe, the apparent consumption of sawnwood and wood-based panels amounted to 131.8 million $m^3$, and the consumption of paper and paperboard amounted to 73.1 million tonnes (MT) in 1995 (UNECE–FAO 1996a). Despite the large forest resources in Europe, the region is a major net importer of forest products accounting for about 27% of global inter-regional trade, which makes it the second largest destination for forest products globally (UNECE–FAO 1996b). The four main markets in Europe that were surveyed for a study (Germany, France, Italy, and the UK) show an apparent consumption of about 56.1 million m$^3$ of sawnwood and wood-based panels (which is about 43% of total apparent consumption in Europe) and about 44.1 MT of paper and paperboard (which is approximately 60% of total apparent consumption in Europe) in 1995 (UNECE–FAO 1997). The most important wood product categories where timber is extensively used are construction, furniture, packaging, and paper, but there is a wide variety of other uses (Oliver 2010).

### 4.6.1 European Business-to-business Markets

A significant amount of empirical data can be found on the European business-to-business market. This is mostly due to an initiative of the World Wide Fund for Nature (WWF) and other environmental NGOs to assist companies in communicating with the market and facilitating communication within cooperation groups. Additionally, of late, some publishing companies have expressed preferential demand for certified forest products, but little data is available. The major characteristics of the market are as follows.

- ***Market size*** The volume of FSC-certified timber bought in Europe in 1996 was estimated to be less than 100 000 m$^3$ of mainly sawnwood. However, this is to some extent an effect of supply restraints, as the FSC trademark itself was only launche d in spring 1996. The current market size controlled by the buyers' group is estimated to be around 9 million m$^3$ of roundwood equivalent (Oliver 2010).
- ***Market growth*** As of September 1997, approximately 1.940 Mha forests were certified in Europe (approximately 1.4% of European forest area). More than 99% of this area was certified according to the FSC system. Assuming annual removals to the order of net annual increment, this would result in roughly 6 million m$^3$ of potential roundwood supply from European forests. The total FSC-certified forest area globally was about 3.1 Mha in September

1997. The volume of certified timber traded in Europe in 1998 was estimated by some experts to amount to 2 million $m^3$ and optimistic estimations speak of 15 million $m^3$ of certified forest product demand in 1998.[4] Markets for certified forest products—the number of buyers groups and/or the number of members of buyers groups—are predicted to grow further in Central Europe. The major potential business market outside the markets in which these groups are active is mainly the paper market where no certified timber is currently traded. The market potential is mainly dependent on the market pull factor—general demand by European companies and especially the already expressed demand by large German publishing companies—and the market push factor—ready supply of products made of certified timber by big suppliers in Scandinavia or Canada. For example, Canadian companies announced plans to certify about 20 Mha of forests that will together produce timber of about 25–30 million $m^3$/year (Rametsteiner and Schwarzbauer 1999).

It is observed that the fashion and the level of economic activity appear to be more important in determining market access than purchasing decisions based on environmental criteria. A slowdown in the construction industry is generally considered to have reduced recent tropical wood imports into France. The current fashion for lighter-coloured furniture has contributed to an increased demand for temperate hardwoods compared to their generally darker tropical counterparts (Upton and Bass 1995).

### 4.6.2 European Consumer Markets: Status

Little is known about the market behaviour of private consumers. Although several private organizations have made surveys or organized test markets for their specific purposes, this data is, in most cases, not available to the public.

Regional variation in market development is an important issue. Private consumer markets with more than the occasional market transactions/purchases of certified timber products can only be found in two European countries—the UK, where it is fairly developed, and to some extent in the Netherlands. Very poor interest in certified

[4] 55th Session of the Timber Committee. Economic Commission for Europe Timber Committee 1997.

wood products seems to prevail in Southern Europe although hardly any data is available (Oliver 2010).

Other product eco-labelling/certification programmes are usually catering to niche markets where the total market share of the different product categories is usually well below 3% even in more attractive markets (v Alvensleben 1992). However, the different structure of the forest sector and the uniqueness of a market support initiative such as the WWF Buyers Groups might result in higher future market shares in forest product markets.

### 4.6.3 Emergence of the Marketplace

Over the past six years, forest certification has grown considerably. As noted earlier, SFI-certified acres have grown from 90 million in 2004 to 181 million in 2009 (101% growth) and from no CoC-certified companies to 807 in 2009 (Fernholz, Howe, Bratkovich, *et al.* 2010).

Globally, since 2004, FSC has expanded the number of certified acres from 125 million to 304 million (143% growth) and from 4000 CoC certificates to 16 235 (305% growth). Since 2007 within North America, FSC has grown from 23 million certified acres in the USA to 31.5 million today (37% growth) and from 1227 CoC certificates to 3542 (189% growth) (Fernholz, Howe, Bratkovich, *et al.* 2010).

In Canada, FSC had 51 million certified acres and 399 CoC certificates in 2007; today, within Canada there are 77 million acres of certified forest and 881 FSC-CoC certificates (51% and 121% growth, respectively). There are some interesting patterns within these numbers. For example, much of the FSC-certified land in the USA is located within two main regions of the country—the Upper Midwest and the Northeast. Just three certificate holders—the Michigan, Wisconsin 22, and Minnesota Departments of Natural Resources—account for 40% of all of the FSC-certified land in the USA. If state lands in Maine, New York, and Pennsylvania (an additional 3.5 million acres) are added to this list, the total rises to 51% of all FSC-certified lands being administered by six certificate holders (Fernholz, Howe, Bratkovich, *et al.* 2010).

Within the SFI programme, the certified land within the USA includes 36% (20 million acres) in southern and 24% (14 million acres) in western states as well as significant acreage in the central and north-eastern regions, that is, 41% and 24 million acres (Fernholz, Howe, Bratkovich, *et al.* 2010).

All over the globe, many state governments face record deficits and budget challenges, and thus decisions about maintaining current forest certification programme commitments will greatly impact the extent of certified forests. In addition to the challenge of declining public-sector budgets and continuing costs of auditing and compliance, continued growth in market demand for certified products represents another challenge to certification. However, two most notable potential areas of market growth are the paper, publishing, and printing sector and the green building sector (Fernholz, Howe, Bratkovich, *et al.* 2010).

## 4.7 STATUS OF FOREST CERTIFICATION IN INDIA

Forest certification in India is still at an early stage; therefore, the nation has not been able to make much use of the benefits of forest certification. So far, India has secured one FSC Forest Management Unit Certificate (644 hectares of rubber plantation in Tamil Nadu) and a few CoC certifications mainly by small and medium companies to meet export demand. Only a few privately owned forest areas have been certified in the past.

Previously, only a few FSC certifications had been carried out in India, including the one for primary forest management in West Bengal.[5] An estimate provided by the International Tropical Timber Organization for the area certified and under sustainable forest management is given in Table 4.1.

Table 4.2 provides the production permanent forest estate (PFE), while Table 4.3 provides the protection PFE. Box 4.1 details the certification of Ballarpur Industries Ltd.

**Table 4.1** Total forest area (range) and area of closed forest ('000 ha)

| *Country* | *Estimated total forest area (2010)* | | *Closed forest* |
|---|---|---|---|
| | *Minimum* | *Maximum* | |
| India (tropical) | 37800 | 37800 | 23100 |

**Source** Country profiles ITTO (2011)

[5] Details available at <www.assets.wwfindia.org/downloads/forest_certification_meeting__background_1.pdf>

**Table 4.2** Production permanent forest estate* ('000 ha)

| *Country* | *Natural* | | | *Plantation* | | |
|---|---|---|---|---|---|---|
| | *Total Area* | *With management plans* | *Certified* | *Total area* | *With management plans* | *Sustainably managed* |
| India | 13 500 | 9 720 | 0 | 32 600 | 8 150 | 4 800 |

*The International Tropical Timber Organization's policies stress the need for countries to establish a permanent forest estate, which is defined as follows: "Land, whether public or private, secured by law and kept under permanent forest cover. This includes land for the production of timber and other forest products, for the protection of soil and water, and for the conservation of biological diversity, as well as land intended to fulfil a combination of these functions"

**Source** Adapted from ITTO (2011)

**Table 4.3** Protection permanent forest estate ('000 ha)

| *Country* | *Area* | | *Allocated for soil and water protection* | | *With management plans* | | *Sustainablg managed* | |
|---|---|---|---|---|---|---|---|---|
| | 2005 | 2010 | 2005 | 2010 | 2005 | 2010 | 2005 | 2010 |
| India | 25600 | 4540 | 0 | 4540 | 0 | 722 | 0 | 722 |

**Box 4.1** Certification of Ballarpur Industries Ltd

The Ballarpur Industries Ltd (BILT), part of the $3 billion Avantha Group, is India's largest manufacturer of writing and printing paper. BILT's subsidiaries include Sabah Forest Industries (SFI), Malaysia's largest pulp and paper company, and BILT Tree Tech Ltd (BTTL), which runs BILT's farm forestry programme in several states in India. BILT has six manufacturing units across India, which give the company geographic coverage over most of the domestic market. BILT has an impressive 85% of the bond paper market and nearly 45% of the hi-bright Maplitho market, besides being India's largest exporter of coated paper. BILT is the first Indian paper company to feature among the global top 100.[#]

Two manufacturing facilities of BILT in Maharashtra have received the international FSC-CoC certification. These certifications apply to BILT's units at Bhigwan in Pune district (SCS-COC-002612) and Ashti in Ghadchiroli district (SCS-COC-002610). BILT is the country's first and only paper manufacturer (virgin pulp based) to be FSC-CoC certified.

With the FSC-CoC certification, BILT has gained membership to the global chain through which wood/fibre sourced from FSC-certified and responsibly managed forests move through the value chain of paper manufacturing. The FSC label would provide a credible link between responsible management of forests and the production, manufacturing, and marketing of wood-based products, including pulp and paper. This certification enables consumers and businesses to make purchasing decisions that benefit people and the environment. FSC-CoC certification has been granted by the Scientific Certification Systems (SCS), an organization accredited by the FSC to perform audits and issue certificates.

BILT claims that strict tracking requirements have been put in place at these two units to ensure that products manufactured and sold as FSC certified come from responsibly managed forests. This certification would also allow the company to manufacture and sell FSC "Pure" and FSC "Mix" products in the domestic and international markets.

[#] Details available at <www.free-press-release.com/news-bilt-is-first-indian-paper-company-to-get-fsc-coc-certification-1250001852.html>

*Source* <www.avanthagroup.com/images/pressRelease pressrelease4809.pdf>

## 4.8 FSC CERTIFICATION IN INDIA

The number of FSC certifications in India has increased from 4 to 43 in the last two years. Several companies are in the pipeline to get themselves certified (Sachin 2010). As on 1 September 2009, India has secured 21 CoC certificates. The demand for FSC-certified paper has increased in India, and this trend is likely to continue in the coming years. A national initiative to develop specific Indian standards of FSC principles and criteria is also in the offing and is likely to be in place shortly.

Some of the companies that have already been granted FSC certification (Sachin 2010) are as follows.

- BILT Graphic Paper Products Ltd
- ITC Ltd
- Thomson Press India Ltd
- Gopson Papers Ltd
- Manipal Press Ltd
- Kraftile International of Santosh Group
- Khanna Paper Mills Ltd
- Repro India Ltd

To facilitate the certification process, various support agencies have begun to open in India; for example, Green Initiatives Certification and Inspection Agency (GICIA) under the banner of SCS has been providing third-party environmental and sustainability certification in South Asia, South East Asia, and the Middle East for more than two years. They have already established offices in Sri Lanka, Dubai, and all cardinal regions in India itself. Earlier certifications issued in India are listed in Table 4.4.

Another notable example is that of ITC, the paper maker of India. Of late, the pulp, paper, and paperboard manufacturing facility of ITC at Bhadrachalam has also been awarded the FSC-CoC certificate. It is the most widely respected certification for responsible fibre sourcing around the world. This certification provides a unique competitive edge to ITC's premier category papers and paperboards in an increasingly environment-conscious world.[6]

[6] Economic Times, 29 March 2010

**Table 4.4** Earlier certifications issued in India

| *Name of industry* | *Certification details* | | | | | *Details of certified products* | | *Current status of certification* |
|---|---|---|---|---|---|---|---|---|
| | *Programme* | *Certifier* | *Certification number* | *Certification type* | *Certification date* | *Product* | *Other details* | |
| Toysin Woodcrafts India Pvt. Ltd | FSC IMO | IMO | IMO-COC-21115 | CoC | 11 January 2001 | Furniture, novelty items, toys | (*Acacia nilotica, Dalbergia sisso*) | Discontinued |
| Assam Bengal Veneer | FSC IMO | IMO | IMO -FM-COC-21023 | CoC | 1 January 2001 | Plantation area-432 acres | *Abies grandis* | Discontinued |
| Proxima Steel forage Pvt. Ltd | FSC | Smart wood | SW-CoC-665 | CoC | 21 January 2002 | Hand construction tools | *Fagus sylvatica* | Continued |

**Source** Forest Certification Resource Centre

The WWF-India and ITC have also joined hands to promote sustainable forest management practices. Since ITC's Paperboards and Specialty Papers Division (ITC PSPD) draws from the corporate commitment to the bottom line of economics and environment and continuously works towards becoming the benchmark for sustainable practices in the paper industry in all areas of its operations, it has signed the agreement of membership with the Global Forest and Trade Network (GFTN). Incidentally, the GFTN is the WWF's initiative to eliminate illegal logging and improve the management of valuable and threatened forests, encourage sustainable forest management practices, and minimize the forest footprint of industries trading in or procuring forest goods.[7]

Following the above said agreement, ITC intends to progressively cover the wood sourced from its farm and social forestry programmes under a credible forest management scheme like FSC. The increasing volume of certified fibre sourced from these plantations will enable ITC to offer discerning customers products labelled under credible CoC certification schemes like that of FSC. However, strategically, it will have to focus on business-to-business products such as packaging boards, graphic application boards, a wide range of communication papers, and consumer range of paper products like paperkraft copier and classmate notebooks marketed by ITC's education and stationery products business.

[7] Details availalde at <www.wwf.org>

# 5

CHAPTER

# Process of Certification of Forests

## 5.1 PROCESS OF CERTIFICATION

The process of certification depends on the certifier's reliability. The certifier (usually an organization) verifies whether timber resources are managed on a sustainable basis or not. In this assessment, the certifier gives equal weight to the health of the entire forest ecosystem, which includes adequate wildlife habitat and watershed protection, and the economic and cultural impact of the operation on local communities. This leads to certification, provided all the qualifications are met.[1] Some aspects of the process of certification are examined in detail as follows.

Forest certification in general consists of four elements: (1) standards (acceptable levels of forest management activity); (2) inspections and audits (objective assessments of whether the standards are fulfilled); (3) chain of custody (CoC) audits (identifying a product's origin and tracking it through the system); and (4) certification (a labelling process that indicates that the previous three elements were met satisfactorily). Box 5.1 elaborates on the process of certification.

Two types of certificates are awarded.

1. Sustainable forest management (SFM) certificates that confirm that forests are sustainably managed.
2. CoC certificates that confirm that forest products are made with wood harvested from sustainably managed forests.

[1] Details available at <www.forestry.about.com/od/certification/Forest_Certification.htm>

**Box 5.1** Process of certification

When forest owners wish to have their forests certified, they make a request to the certification body of their choice (depending on various considerations). The application for certification usually entails an application fee or membership in an organization. Auditors then visit the forest and compare what they see in owner records and on the ground with the standards for certification set by the certifying body. For third-party certification, the auditors are independent of the company or organization being certified so as to avoid conflict of interest. Certification is given when the auditors' report indicates that there are no non-compliances. If there are instances of non-compliance, the landowner is given a fixed amount of time to address the issues before certification may be granted. Periodic or annual audits are required by the certifying body. The landowner must pay not only the application fee but also the costs associated with the audit. Hence, for small landowners, the cost per certified forest acre can be prohibitively high.

*Source* Fischer, Aguilar, Jawahar, *et al.* (2005)

When the certification scheme includes the possibility for making claims on products made with wood from a certified forest, there must also be a process in place for controlling these claims (Handford and Nussbaum 2001).

This market- and consumer-oriented approach of forest certification has developed a high level of acceptance among some private and public forest managers (Bettinger, Boston, Siry, *et al.* 2009).

CoC is a process for tracing products from the forest through the production process to the final product. It is a means of controlling all labels and claims made to ensure that they are clear and accurate. To achieve this, three things are needed in a certification process (Handford and Nussbaum 2001).

1. *Standard* Any forest certification scheme needs a standard that lays out the level of management, which must be implemented by the forest enterprise.
2. *Certification process* A defined method must be followed by the certification body for assessing whether or not a particular enterprise has meet the standard.

3. *Accreditation* There must be a process in place to assess each certifying body, which wishes to issue certificates for a particular standard to ensure that they are competent and credible. This process must include the power to prevent incompetent authorities from issuing certificates (Handford and Nussbaum 2001).

To emphasize how essential each of these elements is to a certification scheme, they are often represented as three sides of a triangle as shown in Figure 5.1.

## 5.2 ELEMENTS OF FOREST CERTIFICATION

The certification of forests has two interconnected elements.

1. *Certification of forest management* Third-party certification of forest management activities typically addresses some or all of the accepted SFM criteria and indicators.
2. *Certification of CoC* CoC is important because wood products pass from stump to processing unit, processing unit to wholesaler, wholesaler to retailer, and then to consumer. Therefore, the consumer must be assured that the wood purchased indeed came from a certified forest. All the following stages are basically adapted from Handford and Nussbaum (2001).

### 5.2.1 Certification of Forest Management

Forest certification is applied at the operational forest management level. It is carried out for a particular forest management unit (FMU), which is defined as an area of forest under a single or common system of forest management. This might be a privately held area of forestland or a series of small areas of forestland owned by

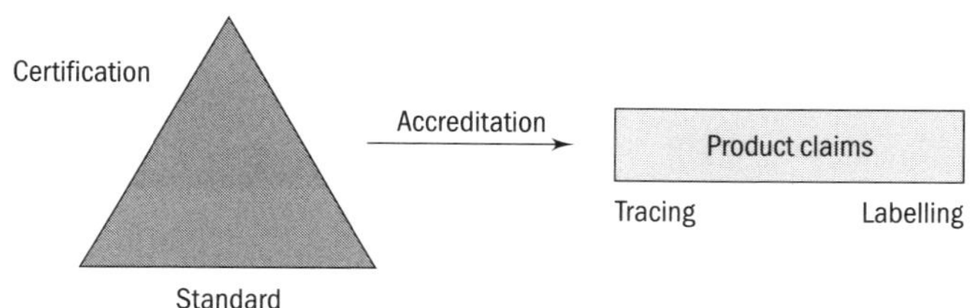


**Figure 5.1** Essential components of a certification scheme
**Source** <www.ogm.gov.tr/yukle/orman_serf.pdf>

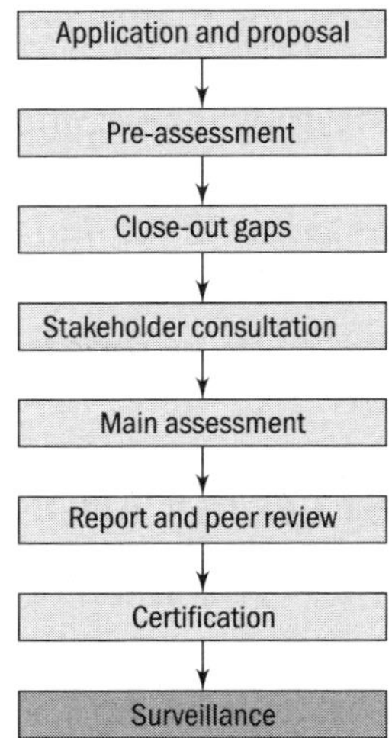


**Figure 5.2** Outline of FSC certification process
**Source** Higman, Mayers, Bass, *et al.* (2006)

different people, but managed under a common system (Higman, Mayers, Bass, *et al.* 2006).

Most of the certification schemes follow a similar pattern. However, differences do occur between the schemes, particularly on the requirements for pre-assessment, stakeholder consultation, peer review, and report writing. The outline of Forest Stewardship Council (FSC) certification process is given in Figure 5.2.

#### 5.2.1.1 *Application and proposal*

The forest enterprise interested in certification completes an application form. Based on the information provided, a financial proposal is prepared with an outline of the certification process and the costs. This is free and provides information on costs without any commitment to continue with the process (Handford and Nussbaum 2001).

#### 5.2.1.2 *Pre-assessment or scoping visit*

Once a forest enterprise has decided to accept a proposal and signed a contract, the first stage in the certification process is a pre-assessment or scoping visit. This is usually a brief visit carried out by one or two people and has three main purposes (Handford and Nussbaum 2001).

1. It allows the certification body to collect information about the forest enterprise and plan for the main assessment.
2. It allows staff in the forest enterprise to meet assessors and learn more about the certification process.
3. It provides an opportunity to discuss the requirements of the standard and to identify any areas where current management does not comply.

#### 5.2.1.3 *Closing out gaps and the decision to proceed*

Any gaps between current management and the performance required by the standard must be closed out. This includes gaps identified as part of pre-assessment and any other gaps identified by the staff of the forest enterprise (Handford and Nussbaum 2001). Once the forest manager is confident that all these gaps have been adequately addressed, the certification body is informed and the main assessment process begins (Higman, Mayers, Bass, *et al.* 2006).

#### 5.2.1.4 *Stakeholder consultation*

Stakeholder consultations aim to obtain peoples' positive and negatives views and comments on the management of the FMU being assessed. The stakeholder may include neighbours and local communities, the forest department and local government, environmental and social non-governmental organizations (NGOs), workers, employees, contractors, and union representatives. Some certification schemes require the certification body to undertake consultation. These bodies are required to contact key stakeholders at least a month before the main field assessment begins. Certification bodies may contact stakeholders by letter or telephone, holding public meetings, advertising in the local paper or any other means, which may appear appropriate. The number of stakeholders contacted and the method used for contacting them will depend on the size and location of the forest and the type of stakeholders being contacted (Handford and Nussbaum 2001). Box 5.2 enumerates the types of eco-labelling.

**Box 5.2** Developing an interim standard for forest management: a case study

The Forest Stewardship Council (FSC) has an international standard in the form of its principles and criteria. However, these are generic requirements and are designed to be globally applicable to all forest types and socio-economic conditions. Prior to use in an individual forest, principles and conditions have to be interpreted for local conditions, which can be done through the development of a national standard, based on the principles and conditions, by a national standard working group (Handford and Nussbaum 2001). Once the standard is complete, it is submitted to the FSC Board for endorsement and subsequently used by certification bodies carrying out any certification in that country (Handford and Nussbaum 2001).

However, this process can take a long time and so many forest enterprises want to be certified before there is a national standard in place. FSC has a process to make sure that this is possible. This is very important because the process of developing a national standard tends to be most rapid in developed countries with many resources, and so if forest enterprises had to wait for the development of a national standard before proceeding with certification, it would tend to discriminate against developing countries. When there is no national standard, the certification body has to develop an interim standard through a process of expert advice and consultation with stakeholders. This interim standard may not be as detailed or precise as the national standard but is sufficiently detailed to allow certification to move ahead. Once a national standard is developed, any forest enterprise that has already been certified on the basis of an interim standard is given 12 months to comply in full with the new national standard. Any enterprise that fails to do this loses its certificate.

*Source* Handford and Nussbaum (2001)

### 5.2.1.5 *Main assessment*

The main assessment is usually carried out by an assessment team over a period of several days (depending on the size and complexity of the forest). The team is appointed by the certification body. Some accreditation bodies specify the qualifications required for assessment of the team leaders and members. The team collects objective evidence to evaluate whether that forest management meets the requirements of the standard, and it identifies any area where evidence requires further scrutiny.

- *Documentation* A range of plans, documents, and records are checked to ensure they meet the requirements of the standard and that adequate records are checked.
- *Field observations* A sample of sites in the forest is visited and checked to see whether the plans and procedures are being implemented. Compliance with legislation, regulations, and local codes of practices is also checked.

#### 5.2.1.6 *Important requirements in the forest certification process*

*Discussion and interviews* The assessment team holds discussions with the managers and staff, and if areas are found where forest management does not comply with the standard, the assessment team records it as a corrective action request (CAR), also known as a condition. This sets out the details of non-compliance and formally requires the forest organization to take action to address the problem. CARs are mainly of two types.

- Major CARs (preconditions) occur when there is complete failure to comply with a requirement of the standard or a systematic failure to implement plans and procedures. Major CARs must be addressed before certification can proceed.
- Minor CARs (conditions) occur when there is partial compliance with a requirement or a non-systematic failure to implement plans or procedures. Certification can still proceed but only on the condition that the issue is addressed within an agreed time.

The main assessment usually concludes with a closing meeting where the team report's back on its findings. Each CAR is discussed and must be signed off by the forest management. The closing meeting provides an opportunity to make clarifications if the team has misunderstood an issue or raised a CAR wrongly because it did not see some important evidence.

Any major CAR identified during the assessment is discussed at the closing meeting. These need to be addressed by the forest management before certification can proceed, and the certification body needs to check that the action taken is adequate. If no major CARs are identified, the assessment team will usually recommend certification.

#### 5.2.1.7 *Report and peer review*

Following the assessment, the team leader, assisted by the team members, will write a report. FSC requires that a summary of the report, including background information and the results of the assessment, be made public; so the report will include both the public summary and the more detailed assessment report. Other certification schemes do not require this type of public summary and only information on the forest or group (size, location, and type) has to be made public (Handford and Nussbaum 2001).

When the report is complete, it will be sent to two or three independent specialists who have been selected as peer reviewers. They are asked to comment on whether, based on the information provided in the report, the assessment seems to be adequate and the findings reasonable. Any issues raised by peer reviewers must be responded to by the certifier and, if appropriate, can lead to new or revised CARs (Handford and Nussbaum 2001).

#### 5.2.1.8 *Certification decision and surveillance*

Once all major CARs (or preconditions) have been adequately addressed and all peer review comments have been responded to, the certification body will finally make a certification decision (Handford and Nussbaum 2001).

The certificate is initially valid for five years but is conditional on the findings from annual surveillance visits, which the certification body will carry out to verify that the forest management is continuing to comply with all the requirements of the standard (Handford and Nussbaum 2001).

In a nut shell, there are two major components that have to be taken care of in forest certification: (1) integration of the social, ecological, and economic factors in management practices and (2) sustainability of forestlands, which can be attained only if the previous integration has been successful.

A typical FSC certification process involves working through the steps of pre-assessment, closing of gaps, forest manager's decision to proceed, stakeholder consultation, development of an interim standard, main assessment, reporting and peer review, and certification decision. Following a positive certification decision, a certificate is issued and annual surveillance visits are undertaken to verify continued compliance with the standard (Handford and Nussbaum 2001).

If seen in the broad context, forest certification is not a single operation but a mix of mechanical and political functions. The relationships between these functions and the responsible bodies are illustrated in Figure 5.3. Furthermore, it is driven by the interests and incentives facing different groups, which range from objectivity to extreme subjectivity (Bass, Thornber, Markopoulos, *et al.* 2001).

## 5.2.2 Certification Process of Chain of Custody

A forest management certificate demonstrates that a forest is being managed in accordance with a defined level of practice. This may be useful for owners and managers of forests seeking investment, government or donor support, or to provide stakeholders with a confirmation of good management practices (Higman, Mayers, Bass, *et al.* 2006). However, if the products of the certified forest are to be sold as products of a well-managed forest, it is essential that there is a mechanism for tracing the products to the forest of origin. This is known as CoC, which can also be defined as the channel through which products are distributed from their origin in the forest to their end users.[2] Most certification schemes have developed elaborate rules for CoC certification that allow products from forests certified under these schemes to be traced to market and labeled as such.

In other words, CoC can be defined as "an unbroken train of acceptability that ensures the physical security of samples, data, and records". As with certification of forest areas, it is important to

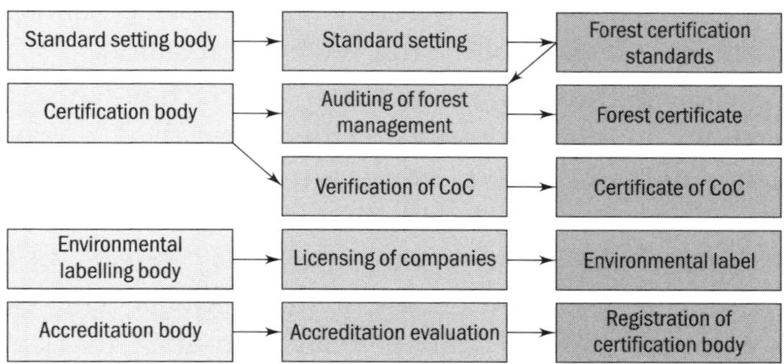


**Figure 5.3** Elements of market-oriented forest certification

[2] Details available at <www.fsc.org>

differentiate between a CoC system (which is installed by various parties in the chain) and CoC assessment (which relates to the activities of the certification body in order to provide a verification of product origin) (Upton and Bass 1995). CoC relies on the following three important components.

1. Segregation keeps certified and uncertified materials physically separate, reducing chances of mixing.
2. Identification ensures that certified materials can be easily recognized at all processing stages, reducing the risk of accidental mixing.
3. Documentation of procedures, work instructions, and record allows the origin of materials entering and leaving a process to be traced.

Taken together these three mechanisms are usually adequate to trace the CoC of materials through a process. Adequate invoicing, records, and procedures are needed to trace certified products during transport, change of ownership, and storage (Higman, Mayers, Bass, *et al.* 2006). For some products, it is possible to assure that a final product is composed entirely of wood derived from a certified forest; these are generally called 100% certified products. However, for more complex CoC, it may be extremely difficult to ensure that all the material input to the process is derived from a certified forest. This can be for the following reasons.

- The total area of certified forest in a region is insufficient to supply enough certified material.
- Certified forests supplying a processor are dispersed at such a distance that the overall environmental impact of transport over long distances would nullify the 100% certification.
- Small- and medium-sized forests, which have not achieved certification, supply a proportion of material to a process and requiring 100% certified wood will exclude them from the market.
- Many certification schemes have, therefore, developed rules for making percentage-based claims about products. This essentially allows manufacturers or retailers to label products that contain less than 100% certified material (Higman, Mayers, Bass, *et al.* 2006). MS ISO Certification is explained in Box 5.3.

**Box 5.3** Need for chain of custody

Logs from a certified forest may pass through numerous manufacturing processes and ownership before reaching the consumer. They may be sawn, peeled, chipped, broken down into fibre, divided into separate loads, combined with other types of wood or fibre, transported, and stored. At any point in this chain, the products of the certified forest could be mixed with those from uncertified forests. Chain of custody certification assesses the controls in place to ensure that mixing does not occur or that it occurs in a controlled and permitted manner.

*Source* <www.fsc.org>

The typical assessment requirements for CoC inspections are illustrated in Figure 5.4. CoC requires that products are identified and segregated and are accompanied by a system of records that can be easily interpreted. CoC must be able to provide physical evidence that the certified product originates from a particular source, requiring secure data capture and a communications system that runs in parallel with and links to the physical evidence (Upton and Bass 1995).

To some extent, there is a trade-off between the need to identify and the need to segregate certified products. An efficient and easily recognizable identification and recording system may reduce the need for segregation. In all cases, the application of CoC systems should use techniques and technologies appropriate to the product. For example, the transport and manufacture of high-value wood products from large logs can justify a sophisticated product identification and recording system related to individual pieces. In contrast, composite wood products using low-quality material—often in partial form—will require a system whose emphasis is on product segregation and batch identification evidence (Upton and Bass 1995).

CoC is a critical element of any certification programme since it provides the link between buyers and sellers from the forest to the point of final sale. For credibility to be maintained, it is important that CoC remains intact throughout, particularly at stages where responsibility for the goods changes. Essentially, CoC is a stock control exercise, which requires transparency and goods to be secure for ease of inspection. The chain itself will consist of a number of links, the number depending on the range of sources, the complexity of the manufacturing process, and the type of market into which the

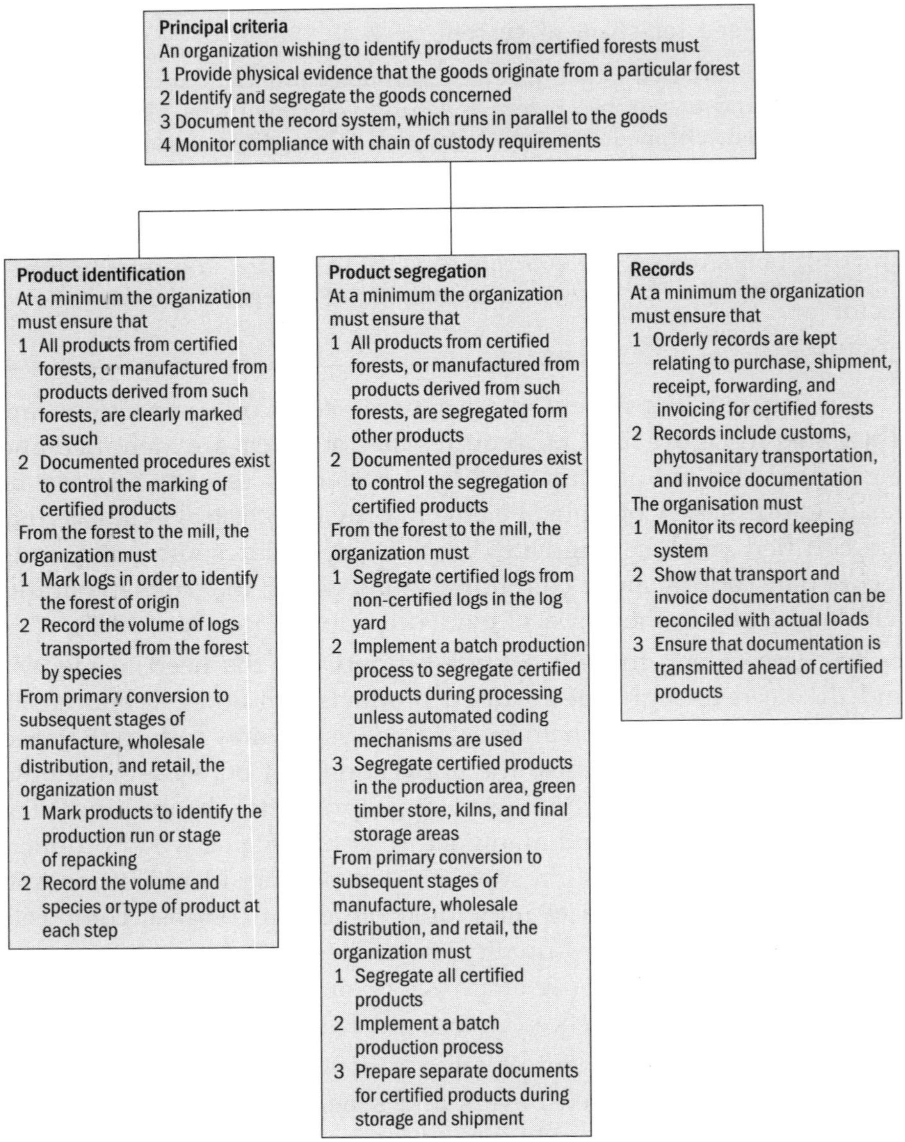

**Principal criteria**
An organization wishing to identify products from certified forests must
1 Provide physical evidence that the goods originate from a particular forest
2 Identify and segregate the goods concerned
3 Document the record system, which runs in parallel to the goods
4 Monitor compliance with chain of custody requirements

**Product identification**
At a minimum the organization must ensure that
1 All products from certified forests, or manufactured from products derived from such forests, are clearly marked as such
2 Documented procedures exist to control the marking of certified products
From the forest to the mill, the organization must
1 Mark logs in order to identify the forest of origin
2 Record the volume of logs transported from the forest by species
From primary conversion to subsequent stages of manufacture, wholesale distribution, and retail, the organization must
1 Mark products to identify the production run or stage of repacking
2 Record the volume and species or type of product at each step

**Product segregation**
At a minimum the organization must ensure that
1 All products from certified forests, or manufactured from products derived from such forests, are segregated form other products
2 Documented procedures exist to control the segregation of certified products
From the forest to the mill, the organization must
1 Segregate certified logs from non-certified logs in the log yard
2 Implement a batch production process to segregate certified products during processing unless automated coding mechanisms are used
3 Segregate certified products in the production area, green timber store, kilns, and final storage areas
From primary conversion to subsequent stages of manufacture, wholesale distribution, and retail, the organization must
1 Segregate all certified products
2 Implement a batch production process
3 Prepare separate documents for certified products during storage and shipment

**Records**
At a minimum the organization must ensure that
1 Orderly records are kept relating to purchase, shipment, receipt, forwarding, and invoicing for certified forests
2 Records include customs, phytosanitary transportation, and invoice documentation
The organisation must
1 Monitor its record keeping system
2 Show that transport and invoice documentation can be reconciled with actual loads
3 Ensure that documentation is transmitted ahead of certified products

**Figure 5.4** Basic CoC requirements
**Source** Upton and Bass (1995)

product is sold (Upton and Bass 1995). Figure 5.4 summarizes the basic requirements of CoC.

Each organization in the chain should establish and maintain procedures appropriate to its scale for identifying individual products or batches from particular sources. Each identification should be unique and recorded. Through the identification and associated records, it should be possible to trace the product to its immediate source, original shipment and/or batch, and certified source of origin. It should also be possible to complete an input–output audit at each organization in the chain. The quantity of certified material bought by the organization should approximate to the amount sold after allowing for processing losses. Usually an appropriate conversion factor and acceptable tolerance limit are agreed upon between the organization in the chain and the certification body (Upton and Bass 1995).

### 5.2.3 Mechanisms for Ensuring Credibility of Assessment Process

One of the best means of achieving credibility of assessment process is effective accreditation of certification bodies. Credibility of the certificate is central to its commercial value and depends almost entirely on the credibility of the process by which certification bodies are accredited.

A local FMU structure, which allows for cost-effective certification, is one of the best means by which certification can be achieved. It is for local FMUs to implement an environmental management system. For certification to be cost effective, the local FMU being certified must be easily assessable. That is it must have a clear and documented management system in place. The level of documentation should be appropriate to the complexity and size of the forest being assessed. An overview of the forest certification programme is explained in Figure 5.5.

Prior to application, the forest manager needs to complete a process of self-assessment to determine if the forestry operation in question is ready for certification. In the absence of an environmental management system, the number of steps taken by the forest owner for implementation of a certification programme is summarized in Annex 6.

## 5.3 COMPARISON OF COC CERTIFICATION SCHEMES

A comparison of two major schemes for CoC is given in Table 5.1. The basic difference is the fact that only the wood certified according to the

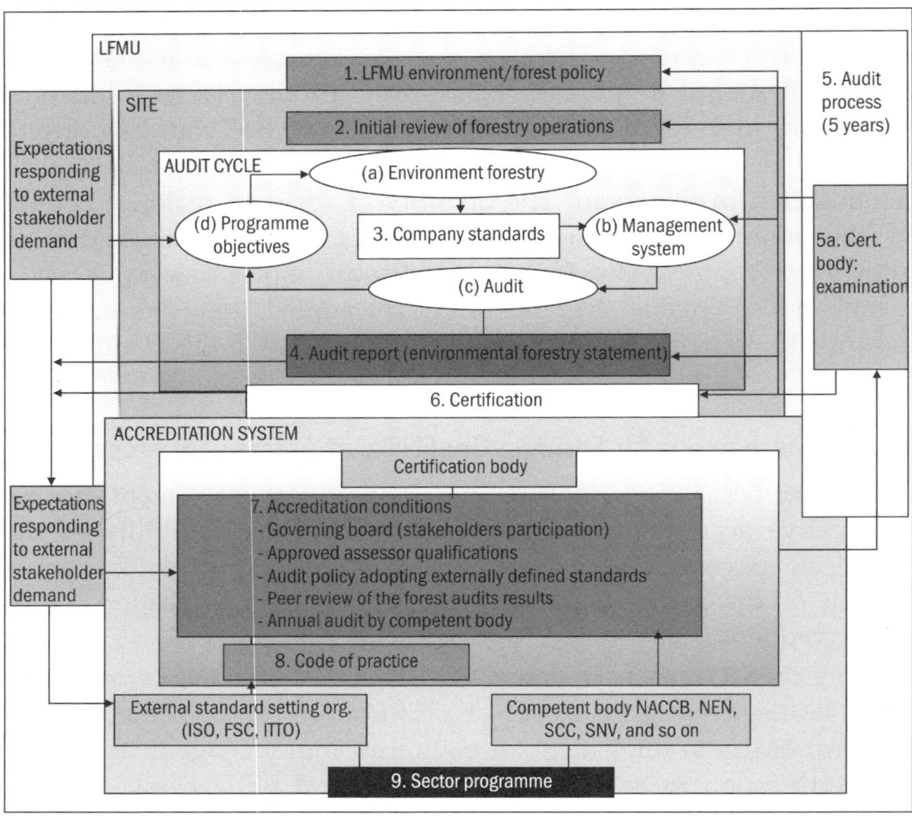


**Figure 5.5** Overview of forest certification process
**Source** Upton and Bass (1995)

requirements of the certification scheme in question is endorsed. Other differences are related to the maximum batch length of production to be monitored, the impartiality of the accreditation body, allowance for recycled fibre, and the minimum content of certified fibre in chip and fibre products. In the latter case, the Programme for the Endorsement of Forest Certification is more demanding than the FSC. The matrix also shows that these elements are common in both schemes, with a few differences. In most cases, verification procedures at the company level are similar across the two schemes (ITTO 2002).

**Table 5.1** A comparison of CoC requirements of PEFC and FSC

| *Feature* | *PEFC* | *FSC* |
|---|---|---|
| Number of products | 1 | 3 |
| Number of categories of materials | 4 | 4 |
| Minimum of average percentage system | + | + |
| Input/output system | + | – |
| Physical separation system | + | + |
| Batch calculation | + | + |
| Maximum batch length (days) | 365 | 60 |
| PEFC-certified wood and products approved | + | – |
| FSC-certified wood and products approved | – | + |
| Verification by an external body | + | + |
| On-product claims | + | + |
| Off-product claims | + | + |
| Additional information on label | + | + |
| Maximum content of recycled fibre (%) | – | 82.5 |
| Minimum content of certified wood fibre (excl. recovered wood/recycled fibre) (%) | | |
| • Collection of solid wood products | 70[a] | 70 |
| • Chip and fibre products | 70[a] | 30[b] |
| • Assembled products | 70[a] | 70 |

FSC – Forest Stewardship Council; PEFC – Programme for the Endorsement of Forest Certification

+ provided; not provided; no limit set

[a] In the minimum average percentage system.

[b] Actually 30%–100% depending on the recycled content.

**Source** <www.fscoax.org>

CoC verification is a technical exercise that could be relatively easily harmonized at the international level. The problem arises from the lack of recognition of forest certificates awarded by other schemes and different minimum requirements for certified raw materials in assembled or mixed products and related claims. The latter could be harmonized without major implications for the credibility of schemes. However, the issue of recognition of certification and accreditation between schemes still needs to be resolved (ITTO 2002).

## 5.4 PITFALLS AND SOLUTIONS TO AN EFFECTIVE CERTIFICATION PROCESS

### 5.4.1 Understanding the Certification Process and Standard

It is important that all parties—both forests managers and stakeholders—understand the certification process and standards that must be met. This ensures that forest managers are working on key issues and effectively preparing themselves for certification. It also ensures that stakeholders are able to follow the logical steps of the process and identify the issues they should raise, which are particularly relevant to the certification (Handford and Nussbaum 2001).

Without a good understanding of the certification process, considerable effort can be wasted on all sides. Providing training in certification requirements at the earliest is worth considering (Handford and Nussbaum 2001).

### 5.4.2 Developing Good Consultation with Stakeholders

The FCS certification process and specific requirements within FCS standards place emphasis on stakeholder consultation. Developing effective stakeholder consultation is fundamental to forest managers being able to progress smoothly towards certification (Handford and Nussbaum 2001).

Poor stakeholder consultation can result in the forest manager having poor information on key matters such as impact on local communities, cultural and historic sites, natural resources, ecosystems, and rare, threatened species. Poor stakeholder consultation increases the chance of major issues being identified during the certification assessment rather than the forest manager being aware of and working on addressing these matters (Handford and Nussbaum 2001).

The following additional points should be considered while developing an approach to consultation.

- Identify all stakeholders and make contact as early as possible. The earlier you start, the more time you will have to get to know people.
- Communicate and develop relationships with stakeholders.
- Understand the requirements and desires of different stakeholders.

- Give meaningful response to issues raised without creating false expectations of what is possible.
- Find appropriate ways to communicate with different stakeholders (face-to-face meetings, public meetings, websites, mail outs, emails, and forest open days). Stakeholder groups often have a few resources and require easy ways to give inputs.
- Build trust. Forest managers need to demonstrate their commitment to meeting SFM requirements under a standard such as FCS and will take account of stakeholder issues. Environmental NGOs and others need to demonstrate that they will not attempt to put companies out of business with unrealistic demands.
- Be prepared to compromise. Be prepared to learn and change where necessary (both forest managers and stakeholders).
- Look at combined efforts between different forest managers/ companies and stakeholders to work together to solve common management issues (for example, reducing chemical use).

# 6
CHAPTER

# Expectations of Stakeholders

## 6.1 STAKEHOLDERS AND THEIR EXPECTATIONS

Having an idea of the expectations of different stakeholders in certification is important. It should be a clear and vital first step in any discussion on forest certification, but it is often ignored. It would be helpful to know the objectives of a particular certification scheme they may like, or why a particular scheme is supported by particular interest groups (Handford and Nussbaum 2001). It is appropriate to have a fair and realistic assessment of the expectations of different stakeholders in the context of "certification of forest". This is very crucial as far as the evolution of an appropriate forest certification is concerned. Stakeholder expectations and actions have positive and/or negative impacts on forests, forest management, and other stakeholders (Higman, Mayers, Bass, *et al.* 2006). Any forestry organization involving outsiders and bringing large-scale investment into an area inevitably raises questions among other stakeholders (Higman, Mayers, Bass, *et al.* 2006). In addition to requiring stakeholders' time and energy, working with stakeholders, incorporating their needs and expectations in forest management, and implementing the wider requirements of sustainable forest management (SFM) create considerable extra work for the forest manager (Higman, Mayers, Bass, *et al.* 2006). Involving large stakeholders in setting targets and agreeing to social objectives involves their time and energy and can leave them with high expectations from the results of their efforts. If these are not realized, it may lead to disillusionment and dissatisfaction with the concept and process. Ways to deal with varied expectations are summarized in Box 6.1.

**Box 6.1** Ways to deal with expectations of stakeholders

The following ways can be adopted to deal with unrealistic expectations of stakeholders.

- Hire/work with people who understand local issues.
- Ensure that the staff who are in contact with local groups and government officials have a realistic picture of the scope of forest operations and the impact on associates. Widely differing views among staff will lead to confused perceptions among stakeholders.
- Respect different positions and keep communication channels open.
- Maximize regular face-to-face contact with stakeholders.
- Ensure continuity of approach and stability of staff in positions involving collaboration with stakeholders.
- Try to solve problems while they are still small.
- Allow adaptation and flexibility in operations involving stakeholders.
- Build alliances with those who understand the real capacities of forest organizations to deliver social benefits.
- Give consultation processes plenty of time and get stakeholders to focus on forestry-related priorities. Stakeholders should be made to understand that forestry cannot be expected to be the main vehicle for local development (an observation shared by most well-respected analysts of forestry today).

In developing initiatives aimed at improved stakeholder benefits, start with small experiments in one area first, ensuring that the stakeholders are part of the plan.

*Source* Higman, Mayers, Bass, *et al.* (2006)

Introductory SFM requirements, including dealing with social issues related to forest management, are likely to be a phased process. Managers may need to discuss, implement, and monitor SFM with the cooperation of stakeholders as a realistic programme of improvement, allowing them sufficient time to develop their abilities without becoming overwhelmed by new demands. A phased approach to implementing SFM requirements can provide a structured framework for implementing a step-by step process (Higman, Mayers, Bass, *et al.* 2006).

Certification is driven by a variety of interests and perceived differently by different stakeholders. For industry and trade, it is an instrument for environmental marketing. For buyers and consumers, it provides information on the impact of products they purchase. For forest owners and managers, it is a tool for gaining market access or market advantage, or perhaps for capturing price premiums. It also serves to demonstrate responsible forest management through independent third-party certification regardless of what the market wants. For the environmental movement, it is a way of influencing how production forests are managed. For governments, it is a soft policy instrument to promote SFM, sustainable consumption patterns, and a variety of other environmental and social goals. For investors, it can help in risk mitigation. Others too may see additional benefits or interests in forest certification (Rametsteiner and Simula 2001).

## 6.2 COMPETING STAKEHOLDER INTERESTS

The causes of forest problems are bound up with the diverging and competing interests of stakeholders such as government, communities in forested areas, employees, investors and insurers, customers and consumers, environmental interest groups, and the general public. Stakeholders have differing and specific interests in the goods and services provided by forest management units (FMUs). Demands on forests from various stakeholders differ; although relationships between stakeholders and strong influences of some groups rather than others do produce a degree of convergence. Some stakeholders, such as environmental groups, are extremely effective at advocating certain perspectives. Incompatibility among diverse stakeholder demands is often a result of the degree to which biodiversity conservation is considered essential to good forestry. Indeed, stakeholder differences are bound up in the ongoing debate aimed at achieving consensus on what is meant by SFM (Upton and Bass 1995).

One thing common to almost all stakeholders—from those with local interests to those with global interests—is the concern about the state of forests, which is greater today than at any other time. Despite this, a large proportion of stakeholders often feel they are left out of the decision-making process. They are confronted with evidence showing continued forest degradation and loss; environmental interest groups have proven themselves to be effective at leading stakeholder opinions and forming new perceptions and demands, through both the market and public policy. Current political systems are often incapable of reconciling stakeholder demands, and because of the long time it takes to become enacted, legislative efforts may not meet current special needs. In addition, the economic and technological power of individual stakeholder groups is now capable of causing a more rapid change. As a result, many stakeholders feel alienated and impotent, and governments are not in an immediate position to help in this situation (Upton and Bass 1995).

Figure 6.1 shows the effect of stakeholder pressures on sustainable forest management

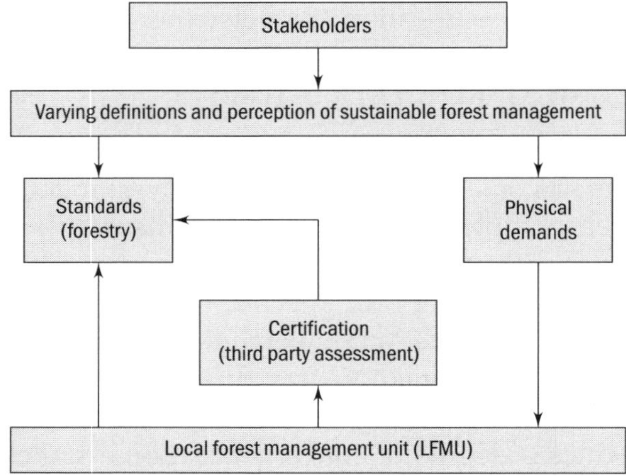


**Figure 6.1** Stakeholder pressures and sustainable forest management

**Notes**

1. The forest management unit (FMU) is under increasing pressure from stakeholders demanding improved management.
2. Stakeholders require that forest management becomes sustainable.
3. Stakeholders have varying perceptions of and definitions for sustainable forest management (SFM).
4. Because of these differences, the forest manager requires clarification on what is meant by SFM.
5. The forest manager requires achievable standards, acceptable to stakeholders.
6. Standard setting requires consensus and trade-off between stakeholders and is difficult as it revolves around a definition of SFM.
7. For this reason, many standard-setting initiatives do not refer to SFM. In this book, we use "quality forestry" as an interim and achievable goal.
8. Stakeholders require credible proof that the FMU has adopted the agreed standards before releasing pressure.
9. Certification is a mechanism where the forest manager can achieve this.

**Source** Upton and Bass (1995)

This chapter attempts to assess varied expectations and requirements for different stakeholders from forest certification concepts and schemes that are emerging with increasing clarity over time and are supposed to be centred on resolving the most urgent environmental problems.

### 6.2.1 Expectations of Retailers

The main objective some retailers would like to achieve by any certification scheme is to obtain support from environmental and social non-governmental organizations (NGOs) and reduce the risk of adverse publicity.

For some industry associations, it is very important that the scheme delivers the industry code of practice for forest management and there is incentive to implement it (Higman, Mayers, Bass, *et al.* 2006).

### 6.2.2 Expectations of Environmental Groups

Environmental groups and forest industry associations have been the earliest proponents of certification as they visualized it as an opportunity to protect environmental values while letting businesses continue to function, with a belief that it allows greater public input into and control of forestland and decisions on use of forest resource. In particular, these groups have led to the creation of major institutions like the Forest Stewardship Council (FSC). Environmental NGOs have supported it and many other certification programmes because it requires independent auditors to verify that certain forest protections are in place, rather than leaving it to companies and governments to make those judgment calls.[1]

### 6.2.3 Expectations of Forest Companies and Forest Industry Associations

Forest companies and forest industry associations may not necessarily support any particular certification programme. But they still value some form of forest certification, mostly because it provides more certainty to their operating climate. Facing incessant demands to change the way they carry out harvesting and other forestry operations, some companies want an environmental stamp of approval that will

---

[1] Details available at <www.proforest.net/objects/publications/verifying_sfm.pdf>

help in creating a better image. For them, certification should show the public that environmental values are being protected by them and that logging and other forestry operations are done in sustainable ways.[2]

### 6.2.4 Expectations of Manufacturers

Certification is also seen by many forest product manufacturers and retailers as a way to capture greater market share in the "eco-certified wood" niche. This niche is often compared to the organic food market.[3]

### 6.2.5 Expectations of People

Public opinion polls have consistently reported that consumers will pay a little more (often called a premium) for products that can be shown to be ecologically friendly as markets evolve.[4] But studies have shown that there is no convincing evidence that a price premium exists for certified timber. Customers are willing to pay more for certified timber, but there is little evidence that they actually are doing so. Important factors affecting this are matters like levels of disposable income, stage of a particular country in the business cycle, and geography. It is more likely that consumer environmental awareness will have a greater impact on the market share of tropical wood compared with wood from temperate areas, and of wood in general against substitute products (Upton and Bass 1995). However, a study conducted by Vlosky, Ozanne, and Fontenot (1999) shows the willingness of customers to pay for green products (Box 6.2).

### 6.2.6 Expectations of Business Groups

Businesses see the need to adapt to new conditions and pursue different customers. But whether forest certification translates into significant price increases and profits is yet to be proved. Hence, in the current phase, the driving reason to achieve certification may simply be that it makes it easier for companies to do business in a world where more people are aware of the environmental, social, and economic damage associated with poor forestry practices.[5]

---

[2] Details available at <www.proforest.net>

[3] Details available at <www.proforest.net>

[4] Details available at <www.proforest.net>

[5] Details available at <www.proforest.net>

**Box 6.2** Market for certified forest products: some evidence

A study conducted by Vlosky, Ozanne, and Fontenot (1999) reveals that there is a positive correlation between the willingness to pay and independent variables. In the model, environmental consciousness, certification, involvement, and perceived importance of certification were considered. The study reveals that there are a cluster of American consumers who have a proclivity to purchase certified wood products. About 16.5 million American citizens have liberal views on the environment and other issues. Besides, at the top there is a second group that sympathizes with the ethical basis of environmentalism but lacks the willingness to make a personal effort in supporting "green products"; this "swing" group represents the key target for green marketers.

*Source* Vlosky, Ozanne, and Fontenot (1999)

Evidence has shown that large producers in industrialized countries are most interested in obtaining certification. Currently, because there is hardly or no price premium for certified timber and because it adds to the cost of forestry management as a whole, certification could be a discouragement for the industry. This is particularly evident in those countries where the most important biodiversity reserves exist but where no proper value is placed on preserving the resource. This can often lead to intensifying those problems that forest certification originally meant to resolve (Cerda and Lira 2002).

Besides all the aforementioned stakeholders, the global community expects and demands more and more ecosystem services from forests (whether public or private).

## 6.3 CUSTOMER EXPECTATIONS

Customer demand is one of the main reasons for seeking certification. For better understanding of their expectation, some of the questions that need to be taken care of are as follows (Higman, Mayers, Bass, *et al.* 2006).

- How many customers and what proportion of sales do they represent?
- Is the proportion likely to grow?
- Are the key customers likely to change suppliers?
- Which certification schemes are they requesting for and what does it cost?

- Will improved sales cover the costs of certification?
- Will the customers assist with the costs of certification and under what conditions?

It is important to be accurate and realistic in assessing the demand for certification. If there is some resistance to certification in an organization, it may be tempting to underestimate demand.

Besides these, it is important to consider what competitors are doing about certification and whether particular certification schemes are being promoted by local trade organizations. Achieving certification may allow an organization to expand its customer base, or it may simply be necessary to maintain the market share it already commands. The organization needs to decide whether customers (current and potential) are consistently demanding certification under a particular scheme, or whether there is sufficient current demand to justify seeking certification under a particular scheme immediately. In addition, it has to find whether there is sufficient likelihood of future demand to justify starting preparations for certification immediately and whether there is no indication that certification is needed now or in the future (Higman, Mayers, Bass, *et al.* 2006).

A survey looking at customers' perception of environmental labels in general revealed that most people (55%) thought the label to be a sort of official approval. Among those who knew that labelled goods do not require official approval, 83% felt they should require it. Overall, the preference was for government departments to take responsibility for approving environmental labels. Nearly 60% consumers favoured government-supervised labelling, and no other organization came close to it. Sixteen per cent would allow manufacturers to issue the labels themselves, and the European Union was named by only 1% (Upton and Bass 1995).

## 6.4 EXPECTATIONS OF INVESTOR, DONOR, AND INSURER

Increasingly investors, donors, and insurers are using certification as a means of reducing their own risks and improving the quality of their investments. Some international banks investing in forestry and agriculture have made commitments to use certifications as a criterion to ensure that their investments do not have negative social and environmental impact. Insurers may also use certification as a means of managing risks on the assumption that a certified, well-managed forest is less likely to be damaged by social unrest or unforeseen environmental impacts. Donors or funders of forestry projects also sometimes use certification as a means of evaluating and monitoring the success of the project or may promote certification as a means of assessing the market (Higman, Mayers, Bass, *et al.* 2006).

7

CHAPTER

# Economics of Forest Certification

## 7.1 BROAD ECONOMICS OF FOREST CERTIFICATION

The process of certification of forests requires a substantial amount of time and energy to document the adherence of management practices to various principles. These have implications and can be divided into direct and indirect costs that apply to both the preparation cost of certification and the cost of undergoing certification.

Many direct and indirect costs of forest certification include the costs incurred to improve forest management and chain of custody (CoC) systems to the levels that are certifiable. These systems include activities such as preparation for audits, actual forest management audits, CoC audits, and yearly monitoring audits. Such direct costs tend to be higher particularly for forest growers located in developing countries. This is because most certifiers/auditors are currently located in Europe and North America and need to be flown in with very high fees and wages compared to local cost structures. Small forest owners in developing countries, who are thus lagging behind required certification standards, need agencies to be available locally (Durst, McKenzie, Brown, *et al.* 2006). The main direct costs include the costs of forest management and CoC certifications. As explained earlier, they tend to be higher for tropical forests than for temperate forests, mainly because most certifiers are located in far-off, temperate countries. This cost differential becomes more apparent as the size of the forest to be certified decreases.

Costs include not only those incurred for changes in forest management and record keeping to achieve forest certification, but also those incurred to cover the expenses of certification inspections, conforming to prescribed changes in forest management, foregone timber revenues, and more stakeholder consultation. On the other hand, benefits can include a better professional reputation, improved worker safety and training, better records, more active public involvement, enhanced knowledge for outreach, better environmental management systems, and perhaps greater timber prices or market access (Cubbage, Richter, Thompson, *et al.* 2003).

### 7.1.1 Direct Costs

The main direct costs of certification of forest management and CoC are payments to certification body. Direct costs of forest management certification are often relatively higher for tropical forests than temperate forests, partly because many certifiers are located in temperate countries and partly because tropical forests are complex both ecologically and socially. Direct costs, which includes those incurred in audits, CoC audits, and yearly monitoring audits, are also relatively higher for small forest management units (FMUs) or forest organizations than for large ones (Upton and Bass 1995). Direct costs tend to be higher particularly for forest growers located in developing countries for reasons cited earlier.

The process of forest certification requires a substantial amount of time and energy to document the adherence to management practices in accordance with various principles. Depending on the work required to meet any documentation requirements of a certification system, the direct certification cost may range from a few cents to several dollars per hectare, depending on the size of a forest and its conditions (Vidal, Kozak, and Cohen 2005). Additional costs may be incurred if substantial changes to management activities are required (Bettinger, Boston, Siry, *et al.* 2009).

Besides this, the direct cost is heavily influenced by the feasibility and ease of assessment by the certification body. This, in turn, is influenced by the extent to which an FMU has a strong and transparent management system in place.

### 7.1.2 Indirect Costs

Indirect costs are related to efforts to upgrade the whole forest management and/or wood processing system to meet the requirements of the certification system desired. Such efforts can be relatively small in cases where forest management was already good enough before certification, which is normally the case in many temperate situations. In contrast, the indirect costs of forest certification become very high if the forest is a long way from good forest stewardship as is the case in many forests in tropical countries (van Dam 2003).

Upton and Bass (1995) have analysed these costs and also those associated with installing the systems that facilitate quality or sustainable forestry. Experience has shown that investment in better planning, documented procedures, training, forest management, and silviculture can yield positive long-term returns. Installing these systems requires increased planning and upfront investment; it also can result in a changed relationship between fixed and variable costs. Fixed costs per unit of output—especially for items like roads—tend to increase. The only way to keep total costs down is to reduce variable costs per unit of output. This can be done by improving the productivity. The level of output at which this trade-off lies will vary from forest to forest, but it often means that average costs at low levels of output tend to be higher with certification than without. This implies that certification of small FMUs is likely to be proportionately more expensive than for large FMUs. In certification, there are economies of scale; hence, small forest owners forming an association for certification can be desirable. The changing relationship between fixed and variable costs is illustrated schematically in Figure 7.1. It can be noted that even where indirect costs increase, the effect on final product prices is lower.

## 7.2 COSTS OF CHAIN OF CUSTODY

Similarly, the cost of CoC certification depends on the management system of the enterprise, particularly control measures and records. Many timber-processing forest enterprises produce both certified and non-certified products, implying additional costs related to the separation of the two types of raw materials and products. Some companies are certified to more than one system (for example, Forest Stewardship Council and ISO 14001), which also has an impact on the costs. As a whole, CoC certification tends to be more costly in tropical countries than in developed countries because forest management systems are less developed in general.

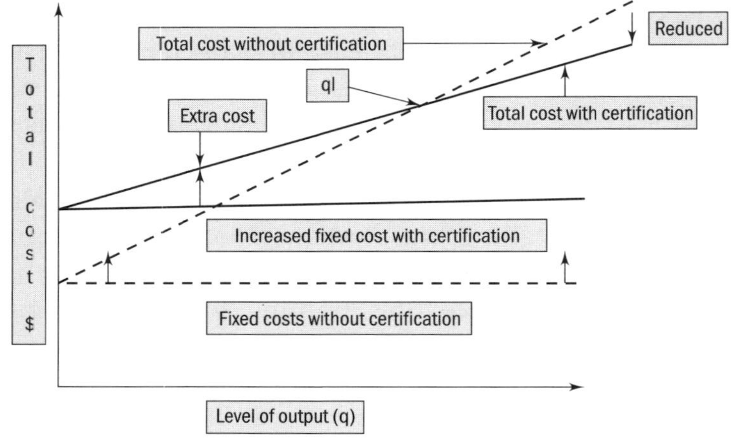


**Figure 7.1** Indirect costs of certification

**Notes**

1. Typically, certification requires an increase in the proportion of fixed to variable costs.
2. Increased fixed costs require a reduction in unit variable costs to contain total cost.
3. *ql* is the level of output where total cost with certification equals total cost without certification. At higher levels of output, the total cost per unit of production is lower with certification. At lower levels of production, the total cost is higher with certification.

**Source** Upton and Bass (1995)

## 7.3 ANALYSIS OF COST

The costs of certification can be segregated under three major headings (van Dam 2003).

1. The costs of certification process itself—what the producer pays to the certification body.
2. The costs implied by meeting all requirements of proper sustainable management to be able to attain a desired level of certification.
3. The costs associated with the maintenance of certification, that is, payment made to the certifier for annual monitoring/verification visits.

Certification costs may vary tremendously, depending on the size, complexity of forest operation, existing management plan, and proximity to a local certifier. In general, costs have ranged anywhere from $0.10/acre up to over $1.00/year (Durst, McKenzie, Brown, *et al.* 2006).

The first of these costs is derived from the certification process itself and is associated with the productive scale of the forest enterprise.

This cost is relatively high and consists largely of fixed costs, which means the smaller the size of the forest area, the higher the relative cost of certification. Box 7.1 expounds on the cost of certification.

| **Box 7.1** Cost of certification |
|---|
| On the basis of studies carried out on six native forests in Central America, De Camino and Alfaro estimated in 1998 a rising initial certification cost of between $0.45/ha (for a 36 hectare plot) and $10.66/ha (for a hectare plot). The annual verification cost was identical in all these cases ($2000), which implies a cost per hectare of $0.55 for 36,000 hectares and $2.66 for 750 hectares (van Dam 2003). |

If the cost associated with initial certification and its maintenance is prohibitive for smaller forest owners and communities, the same logic holds good for the second cost also, which is derived from all kinds of investments necessary to meet proper management standards. In this case, not only greater costs of distance or accessibility are added to those of scale, but also costs get increased by the engagement of professionals to draw up and implement forest management plans and meet the high standards demanded for safety, storage, transportation, and management of agrochemicals and for payment of social stipends set by local labour laws. All this makes certification unviable for the owners of small forests and for indigenous and rural communities, who often lack entrepreneurial capacity and the capital necessary to achieve this sort of management (van Dam 2003).

## 7.4 CONSEQUENCES OF HIGH COST

Certain studies (Gan 2005) show that certification can increase production costs by 5%–25% and as a consequence may negatively impact the world's forestry output, creating a hike in global certified forest product prices. According to Gan (2005), the regions that would suffer most from global certification will not be major timber-producing regions, but major net importers of forest products like East Asia. Therefore, cost of certification is among the most important issues that have to be addressed if forest certification has to spread to become a widespread movement.

The majority of certified forests at present are industrial forests and plantations. In countries like India, a small proportion of privately owned small forests exist where ownership lies with the individual, community or institution. The issue of certifying such small forests has been given less attention by most of the leading forest certification schemes over the years. As a result, these groups are generally under-represented in certification exercises. The main concern of private smallholders is that the needs and circumstances of small forest growers as well as their regional variations in sustainable forest management are not easily accommodated in the dominant certification schemes (Kanowski, Sinclair, and Freeman 1999). Some certification programmes have recently targeted certifying forests owned by small forest holders and taken measures to assist small landowners in meeting challenges through introducing programmes such as group certification to bring down the cost of certification (Cubbage, Richter, Thompson, *et al.* 2003).

Lower cost option for small forest owners through group certification is described in Box 7.2.

**Box 7.2** Group certification: lower cost option for small forest owners

The cost of certification has always been a barrier for forest owners in small countries and particularly for small forest owners. The certifying schemes have come out with a solution in the shape of group certification. Here small forest owners can apply and attempt to get aggregated for the purpose of "certification". Through this aggregated form, the cost of certification gets substantially reduced. Thus it becomes economical for both the certifying agency and small forest owners.

Changes in forest management practices involve maintaining separate inventories of certified and non-certified products (which leads to increased material handling cost), the costs of tracking a certified product through the system to the customer, and the costs associated with becoming and remaining certified. These are some additional costs associated with certification. This sometimes proves to be prohibitively high especially for forest owners (and particularly smallholders) in developing countries (Durst, McKenzie, Brown, *et al.* 2006).

In a study, three participating organizations in the Southern Center for Sustainable Forests—North Carolina State University,

Duke University, and North Carolina DENR Division of Forest Resources—recently received both Forest Stewardship Council (FSC) and Sustainable Forestry Initiative (SFI) certification for their forests. Forest certification required costs of changed management either to meet new standards, as in case of SFI, or to meet new conditions, as in case of FSC. These costs consisted of (1) added employee time and expenses to keep records and train staff; (2) more time and funds needed for changes in management practices; and (3) opportunity costs of managing the forest in an environmentally sensitive manner that would yield less income than could be achieved by profit maximization alone (Cubbage, Richter, Thompson, *et al.* 2003). A summary of information about the direct and indirect costs and benefits of achieving and retaining certification for the given forests is given in Box 7.3.

**Box 7.3** Costs and benefits of forest certification: a case study

A study revealed that the costs to prepare for audits were usually more than the costs to pay audit firms. Total costs depended on the time spent to prepare for inspections and the number of personnel involved in the certification process. Average costs depended on these factors divided by the number of acres being certified. Thus the smaller North Carolina State University forests were more expensive. Specifically, the preparation costs of the Division of Forest Resources were only $0.08/acre for Forest Stewardship Council (FSC) certification and $0.18 for Sustainable Forestry Initiative (SFI) certification. Duke's average preparation costs were $1.21/acre for FSC certification and $3.08/acre for SFI certification. North Carolina State University's costs were $3.73/acre for FSC certification and $7.90/acre for SFI certification. The costs for each institution were estimated based on the number of hours that personnel spent on certification efforts. Each organization tracked the time devoted to each certification process. Time was divided into the following categories: (1) preliminary meetings and general preparation; (2) pre-audit meetings and preparation; (3) document preparation and collection of evidence; (4) office visits by auditors; (5) post-audit work; and (6) report analysis and response. The opportunity costs that foresters and staff incurred on certification were based on a fixed rate of $50/h multiplied by the number of hours spent preparing for or responding to certification-related queries (Table A). There were probably small operating costs of a few hundred dollars to $2000 for copying, notebooks, mailing, use of vehicles during field inspections, and meals.

*Contd...*

**Box 7.3** *Contd...*

**Table A** Estimated total direct SFI and FSC certification costs by the organization (in $)

| Cost/certification system | DFR | Duke | NCSU | Total |
|---|---|---|---|---|
| Acres | 42000 | 8000 | 4500 | 54500 |
| **Inspection cost** | | | | |
| FSC | NA | NA | NA | 78450 |
| SFI | 11526 | 8839 | 16960 | 37325 |
| **Preparation costs** | | | | |
| FSC | 3,350 | 9750 | 16787 | 29887 |
| SFI | 5850 | 24600 | 35562 | 66012 |
| **Total costs** | | | | |
| FSC | NA | NA | NA | 108337 |
| SFI | 17376 | 33439 | 41960 | 92755 |
| **Inspection costs/acre** | | | | |
| FSC (54,500 acre) | NA | NA | NA | 1.44 |
| SFI (44,500 acre) | 0.36 | 1.10 | 3.77 | 0.82 |
| **Preparation costs/acre** | | | | |
| FSC | 0.08 | 1.21 | 3.73 | 0.55 |
| SFI | 0.18 | 3.08 | 7.90 | 1.48 |
| **Total costs/acre** | | | | |
| FSC | NA | NA | NA | 1.99 |
| SFI | 0.54 | 4.18 | 9.32 | 2.30 |

DFR – Division of Forest Resources; FSC – Forest Stewardship Council; NCSU – North Carolina State University; SFI – Sustainable Forestry Initiative

**Note**

As noted, costs of FSC inspections were more in total than that of SFI inspections. However, one should not generalize that this will always be the case. Auditing costs will depend on the forest, the organization being certified, and the firm performing the auditing. The costs per acre are much higher for small forests. In addition, scrutiny of all operations is somewhat more intense for forests that we did not perceive as an advantage at the North Carolina State University. FSC costs could not be estimated for each institution. The total direct costs for all organizations for certification inspections and preparation averaged $1.99/acre for FSC and $2.30/acre for SFI. These costs are for the initial certification. The required annual revisits by the FSC team will add annual management costs, as will the required management changes. Major site visits at the time of recertification (five years for FSC and three years for SFI) will again require costs similar to or more than those calculated above.

**Source** Cubbage, Richter, Thompson, *et al.* (2003)

To conclude, one can infer that the benefits of certification are mostly indirect, although surely valuable. They can include a better professional image, improved worker safety and training, better records, more active public involvement, better morale among forest managers, enhanced knowledge for outreach and extension, better environmental management systems, and perhaps greater timber prices or access to more markets. Direct certification costs for each organization ranged from less than $1 to almost $10/acre per system for the initial efforts and may cost another dollar or two per acre on an annual basis (Cubbage, Richter, Thompson, *et al.* 2003).

## 7.5 OVERALL ECONOMICS OF FOREST CERTIFICATION

The market demand for green products is not even, and there is a great difference between consumer awareness and the willingness to act on this awareness. This also varies from country to country. Countries outside Europe and the USA are experiencing very little environmental pressure. Within Europe, the UK, Germany, and the Netherlands demonstrate a high level of awareness compared to France, Italy, and Portugal, where environmental concerns are low, although increasing (Upton and Bass 1995).

In general, certification costs tend to be much higher for primary producers than for processors. On the other hand, the benefits of certification, which relate mainly to market access, tend to be realized by actors further down the supply chain. Therefore, at present, the main winners from forest certification appear to be far from the forest, particularly in case of tropical forests. This makes the economic scenario for forest growers (particularly smaller ones) in tropical countries not very favourable at present. However, as green awareness increases across the globe, the economics of forest certification may turn out to be viable.

8

CHAPTER

# Addressing Sustainability

## 8.1 OBJECTIVES OF FOREST CERTIFICATION AND STATUS OF CURRENT ACHIEVEMENT

Forest certification as a concept and process has now evolved into a major worldwide movement and is quite well documented in the literature and on the web. It has been the subject of numerous worldwide deliberations in conferences and seminars and is getting a lot of global exposure. However, the final report card on forest certification is not out as yet and thus the effectiveness of the process is still under debate.[1]

Although interest in forest certification has been wavering since its inception, it remains a cornerstone of forest policies. But what is the performance today, more than two decades after the first forest certificate was issued? It should now be possible to evaluate the effectiveness of certified forest management by comparing the conservation performance of certified forests with non-certified forests (ETFRN 2010).

The question whether forest certification has been effective in achieving its different objectives, including conservation of ecology, is difficult to answer at this stage because there is still not much evidence available. A literature review reveals the difficulty of providing a clear answer. In the first place, in most certified forests, the data needed to assess the effects of management on biodiversity are not being systematically collected. Data from non-certified forests, which are needed for comparison, are even more difficult to find. The scientific community has not yet risen to the challenge of providing evidence of the effects of certified forest management on a comprehensive scale. Also studies focusing on different species use different protocols and do not address the large temporal and spatial scales covered by

[1] Details available at <www.forestry.about.com/od/certification/Forest_Certification.htm>

tropical production forests (ETFRN 2010). But there are indications that independent audits are providing an incentive for improving forest management (Rametsteiner and Simula 2001). Gullison (2003) has concluded that forest owners who undergo certification are more likely to improve management with respect to the value of managed forests for biodiversity conservation.

Many issues have not been adequately analysed in "certification", which will determine its effectiveness. The overenthusiasm of forest certification promoters to establish as many certified forests or initiatives on the ground as possible sometimes causes the basic objectives for pursuing certification to be lost. Certification, in some places, has become a goal in itself rather than the facilitator of adequate forest management practices. This may force many forest certifiers to develop mechanical checklists that can be used to assess all types of forest (Vogt, Larson, Gordon, *et al.* 2000).

The effectiveness of certification derives from the following four main impacts (Bass, Thornber, Markopoulos, *et al.* 2001).

1. *Stretch demanded by certification criteria compared to commonly enforced regulatory requirements for forest management* The commonest effect is that certifiers are requiring producers to meet all current legal requirements that they might normally not bother to meet. But new national standards are beginning to raise the level of all legislation or at least make it more practical and a better match with stakeholders' concerns.
2. *Extent of certified forest area under different forest and producer types* Certification is being applied routinely in many temperate and boreal forests. Much of the discussions and developments concern the competition and equity among producers who are just above or below the threshold of acceptable forest management as defined by the Forest Stewardship Council (FSC) in particular. However, this will require a stepwise approach in terms of standards and incentives to help improve poorer producers in many tropical contexts, such as in the Indonesian scheme.
3. *Demonstration effect of certified forests on other forests* Certification has rather limited influence on the management of non-certified forestland. Such influence has often been constrained by external support to certified enterprises, making them appear as atypical to neighbouring forest managers. Some certifiers are developing ways to make demonstration a more overt aim of their schemes.

4. *Changes in consumption patterns towards sustainability* Forest and trade networks have always been the key, although their scope is still rather limited and they are only just making inroads into markets for paper products and construction.

Some emerging useful trends are bringing costs down—group certification schemes, use of local inspectors, price competition among certification schemes and inspectors, streamlining audit procedures, focusing on outputs or environmental/social outcomes to permit flexibility for producers, and national working groups developing local standards, which allows resources to be focused on what really matters. However, although it is beginning to be exercised for some unique non-market roles, certification shows few signs of being more efficient than other means in its non-market purposes.

## 8.2 SOME OBVIOUS EFFECTS OF CERTIFICATION

Not much work has been done with regard to a clear quantification of the impact of green certification/different aspects of forest certification on conservation. Hence, there is a need for an effective monitoring system that clearly documents the gains and losses to conservation and other aspects from certified and uncertified forests (Nussbaum and Simula 2004).

Nussbaum and Simula (2004) have provided a series of case studies indicating potential benefits of certification to biodiversity conservation. They have concluded that the improvement in conservation of biodiversity appears to be a consistent benefit of forest certification as evidenced by increased protection of representative ecosystems and rare, threatened or endangered species. But there is little data available on which rigorous assessment of its effectiveness can be based. Loehle, Wigley, Rutzmoser, *et al.* (2005) evaluated associations between a set of landscape structure metrics used in certification programmes and bird species richness. Their work suggested that indicators addressing aspects of landscape structure may be useful in evaluating potential contributions to biodiversity conservation. But the extent to which certification has led to improved ecosystem functions is still unclear. Nonetheless, certification represents a pragmatic international system of private governance that could lead to increased levels of conservation (Nussbaum and Simula 2004).

## 8.3 MEASURING THE IMPACT OF CERTIFICATION

Any analysis of on-the-ground impact of forest certification is quite challenging for two major reasons.

1. *Diversity and range of certification standards* Different certification standards such as FSC, Sustainable Forestry Initiative (SFI), American Tree Farm System (ATFS), and Green Tag differ in terms of content and certification process. Moreover, perhaps most importantly, nuances among standards change over time, as they incorporate new information or sometimes compete with each other for legitimacy in the marketplace (Cashore and McDermott 2004).

- Everything else being equal, the most progressive firms are likely to pursue certification first (Hayward and Vertinsky 1999). Unfortunately, forest units whose practices must be improved by certification may be the least likely to join, thus defeating the very purpose of certification.

## 8.4 ISSUES RELATED TO CERTIFICATION STANDARDS

### 8.4.1 Inflexibility of Standards

The fact that only a few natural tropical forests have been certified over the past is an indicator of the inflexibility of standards of performance. Tropical forests, where efforts to implement sustainable forest management (SFM) are recent and often far from definitive, are in a disadvantageous position because certification standards tend to focus on the end results of SFM practices. Certification standards do not recognize successive steps to go through when moving towards SFM. In addition, certification standards that specify particular types of inputs and technologies can impose a greater cost burden in tropical situations than in developed countries or even exclude producers without access to required inputs (Markopoulos 1998).

### 8.4.2 Lack of Recognition of Broader Local Land-use Issues

Certification primarily focuses on forest management units and often fails to take into account other surrounding, related land-use issues such as development of agriculture, which can have a significant impact on forests. For example, comprehensive land-use planning at the landscape or regional level may include delineating permanent forests and designating forested areas where development of agriculture will be allowed. In such a case, only permanent forests would be concerned by certification, which would not recognize efforts to sustain land-use practices on a larger scale. Certification is not able to address the root causes of deforestation, nor is this its purpose (Atyi and Simula 2002).

Forest certification protocol has so far tried to ensure or led to private governance of forest management and thus provides social license for forest managers to continue their forestry business. But incentives to forest owners for certification in tropical forests are not adequate and apparent, given the high costs associated with certification. Although the benefits of certification to different conservation aspects are not known, the generally accepted assumption is that certified forests should do more for conservation and other related aspects than uncertified forests. Sufficient statistics are needed to critically confirm or reject this assumption (Cerda and Lira 2002).

Although there are no statistics to assess the magnitude of the effects of forest certification in the market for forestry products, it is estimated that as of December 2001, around 10% of the world's productive forests and 8% of all lumber sold have had some sort of internationally recognized forest certification.[2] On examining the evolution of areas certified by one of the main promoters of certification, FSC, one can appreciate that there has been an upward trend in forest certification around the world that has come about in less than one decade (Cerda and Lira 2002).

Judging by the literature on the subject (Bass and Simula 1999; Cerda 2000; Upton and Bass 1995), one begins to identify as many as 10 objectives that forest certification attempted to tackle. Some of them are social in nature, others financial, and the majority environmental. For the purpose of this chapter, the focus will be on those closely related to the last-mentioned dimension, that is, sustainability. This issue can be summed up as (1) reducing and/or reversing the deforestation process; (2) improving forest-based management and encouraging multifunction of forests; and (3) reducing environmental impact on forestry production.

The aim of the chapter is to examine the effectiveness of forest certification and how the aims mentioned above have been achieved. For this information available on the subject and the experiences at hand will be used, especially those of important certification agencies, which have the support of environmental non-governmental organizations.

---

[2] Details available at <www.fscoax.org; www.pefc.org; www.fao.org>

## 8.5 FOREST CERTIFICATION: DOES IT REALLY CURB DEFORESTATION?

The logic behind certification is hindering or minimizing the access of those producers to forestry markets who do not follow sustainable practices, especially those who are not committed to ensuring the continuity of forestry resources once harvesting has been completed. Certification achieves this aim, but this has to be proved by the following outcomes.

- The market for forestry products regulated by certification is a significant fraction of the market for the entire forestry production.
- Forest producers who currently follow unsustainable practices see certification as a more profitable option than the one they have been following and thus decide to change their modus operandi.
- Forest producers do not have any more profitable options open to them than forestry activities.

None of these three conditions are difficult to verify in practice. In the first place, only 20% of wood production accounts for international sales (Bass and Simula 1999). This situation is exacerbated by the fact that over half of the world's wood production is used locally as fuel, something which is not associated either with formal markets or with regulations of any type whatsoever.[3]

## 8.6 DOES FOREST CERTIFICATION REALLY IMPROVE FOREST MANAGEMENT?

If one examines only the direct effects of the introduction of forest certification, the answer is positive. The quality of forest management by certified producers is better than what it was prior to becoming certified, either because they adopted ISO 14001 type environmental management systems or because they complied with certain performance standards—FSC, Programme for the Endorsement of Forest Certification (PEFC), SFI or some other (Bass and Simula 1999). Despite this, generally large conglomerates that already follow proper forestry management practices or, to a lesser extent, marginal producers who can receive cash incentives from international organizations are most interested in certification (Markopoulos 1998, 1999a,b).

[3] Details available at <www.fao.org>

Of late, certification is also considered an initiative to motivate industrial companies producing and releasing $CO_2$ into the atmosphere to balance their industrial outputs of carbon by sequestering an equivalent amount of carbon in forest biomass. Thus forest certification can be a tool to ensure that these goals could also be satisfied at the ground level in the most efficient and effective manner (Vogt, Larson, Gordon, *et al.* 2000).

The nonexistence of price premiums for certified wood often leads to the dichotomic option of either acceptance or rejection in international wood markets, if there is persistent demand for the present requirements regarding certification. This means those in favour of this tool might compel non-certified producers to modify their practices to be able to gain access to international markets (Simula 1999). However, such an option would only be realistic if the alternative option of not becoming certified were to be proved less profitable, but there is no evidence to back this up.[4] The "natural" danger is that marginal producers who see the export markets closed for them will try to cash in their assets with little heed to sustainability or remain oriented to local markets where performance standards are low (Cerda and Lira 2002).

Moreover, pressure regarding the different certification options is oriented to ensuring multifunctional forests but on a minimal spatial scale, which is not only inefficient (Vincent and Binkley 1993) but also impractical in many cases for marginal producers.

## 8.7 DOES FOREST CERTIFICATION REDUCE THE ENVIRONMENTAL IMPACT OF FORESTRY?

Forest certification can reduce the environmental impact of forestry depending on whether one concentrates only on the direct effects of certification or if one examines its indirect consequences. For those who have obtained certification and/or are still going about obtaining it, the question can be answered affirmatively. However, on a global perspective, the answer is negative because of the following reasons (Cerda and Lira 2002).

- Certification, as known until now, is unfair to marginal producers because only large producers or those close to sensitive markets are able to bear the high transaction costs it implies.

[4] Details available at <www.fao.org/forestry/fop/fopw/gfss/gfsswp/3/HTTOC.HTM>

- Certification is unfair to forestry activities because of the increase in costs derived from certification practices. Thus it discourages wood production as opposed to substitute products generally obtained from non-renewable resources and call for energy requirements several times greater (Koch 1992; Simula 1999).
- Certification brings perverse incentives into play because while requiring environmental performance levels not differentiated in practice, it induces marginal producers to embark on unsustainable and short-term practices.

A case where certification on plantations had positive impact is given in Box 8.1. Box 8.2 discusses how certification is a better option for workers in Brazil.

**Box 8.1** Case of positive impact of certification on plantations

***South Africa: wetland management***

The certification process highlighted a number of common issues (major plantation) companies had to address, including the following (Atyi and Simula 2002).

*Water monitoring* The main environmental issue associated with forestry in South Africa is its impact on water sources. Three big companies SAFCOL, Mondi, and Sappi realized that this was a common issue and established a joint water monitoring strategy and shared methodology. This system is beginning to show results.

*Riparian zones* Mondi managers in particular highlighted river course management as an area that benefited from scrutiny of certifications. A delineation protocol has now been developed with stakeholders, which defines the location of wetlands in the landscape. Several industry representatives agree that the most significant physical impact on plantations of the improved practices encouraged by certification is due to the criteria relating to watercourse management.

*Road building and maintenance* Forestry roads are often neglected and serve as a continuing source of erosion and pollution of watercourses. When one company was issued with a corrective action request on road maintenance, it responded by appointing a "roads champion" who developed revised road building and maintenance guidelines and ran a training course for company employees. In the certification process, another recognized that it had too many roads in its plantations on an average and is now managing a programme of planting grass over some roads.

*Source* Clarke (2000)

**Box 8.2** Brazil: a better deal for workers through certification

The main outcome of Forest Stewardship Council (FSC) certification in relation to plantations seems to be that certified plantations pay their employees according to Brazilian law and often above the minimum salary. As a result of certification, plantation employees also have long-term employment contracts as well as enjoy health and other social benefits. This is not always the case in non-certified plantations and certainly not the case in illegal operations. Non-certified plantations take less care of their employees. Therefore, areas under FSC certification contribute to poverty alleviation and biodiversity conservation. FSC does not seem to have an impact on large-scale developments, especially illegal deforestation. FSC seems to hamper a negative development; where an FSC certificate has been issued, the forest is still standing. However, there is still ample space to expand in the Amazon.

*Analysis of subcontractors' working conditions in FSC-certified Madeireira Itacoatiara Ltd*

The study found that FSC certification led to major changes in the way companies treat their workers.

- Improved living conditions in remote logging camps (such as clean water supply, kitchens, television and other entertainment, safe transport, regular medical check-ups).
- Job stability and work contracts (such as unemployment compensation, paid vacation, 13th salary); FSC principles are applied equally to all workers, either hired directly or through another company, carrying out activities in the forest management unit.
- All workers wear protective equipment supplied by the company, which are in good condition and adequate for their specific activities; they are properly trained to carry out forest activities safely and efficiently.

Up until 1996, the Municipality of Itacoatiara in the state of Amazon was considered the most vulnerable place in Latin America having the highest rate of fatal and major accidents related to forest activities. Every year, two or three people died working for Madeireira Itacoatiara Ltd (MIL), the main local forest company. Since its certification in 1997, the company had one fatal accident caused by a falling dead tree.

*Source* de Azevedo and de Freitas (2003)

9

CHAPTER

# Certification of Forests and Issues of Environmental Conservation

## 9.1 OBJECTIVES OF CERTIFICATION REVISITED

The prime objective of "certifying a forest" has always been the maintenance of ecological integrity. All related aspects are addressed by having sound management practices, ensuring ecological integrity, and leading to sustainable forestry. Environmental conservation issues in certified forests are important. In this chapter, the broad meaning of environmental conservation has been discussed and at times biodiversity or other aspects have been taken up as interchangeable terms.

People from all walks of life value biodiversity for spiritual, aesthetic, recreational, and other cultural reasons, and that is why it forms a very important attribute to any sustainable forest.

Biodiversity forms the foundation of the vast range of ecosystem services that contribute to human well-being (*Millennium Ecosystem Assessment 2005*). Biodiversity in plants, animals, and microbes ensures the production of food, clean water, fuels, and medicine. Diverse ecosystems capture and convert energy into a variety of materials, process waste, drive global biochemical cycles, and regulate the world's climate. In general, they affect the strength and capacity to provide essential goods and services necessary for the well-being and prosperity of people in both developed and developing countries (Martens, Rotmans, and de Groot 2003; MEA 2005).

### 9.1.1 Biodiversity and Sustainable Forest Management

At least 40% of the world's economy and 80% of the needs of the poor are derived from biological resources.[1] Since biodiversity origins

[1] Details available at <www.cbd.int>

mainly in forests and wild areas, their conservation assumes significant importance in the context of sustainable forest management (SFM). Realizing this, many SFM standards across the globe assign adequate importance to "biodiversity conservation" through appropriate criteria and indicators. For example, India's draft national criteria and indicators for SFM try to address biodiversity issues either directly or indirectly. The criteria and indicators of the Bhopal India Process are given in Table 9.1. Annex 6 summarizes how different certification schemes address the biodiversity component.

## 9.2 CERTIFICATION: A TOOL FOR CONSERVATION

Forest certification is widely seen as an important tool for ensuring conservation of forests. In fact, during the 1990s, concern about the loss of biodiversity in logged forests was one of the key drivers behind the emergence of forest certification. It emanated from the premise that production forests could play an important role in conserving nature by following standard practices of forest management that incorporates preventive and curative aspects to take care of the effects of logging and other forest management activities on biodiversity (Gullison 2003).

For this, a comprehensive strategy for conserving the world's forests has to be implemented, which could and should be in a position to minimize the ill effects that industrial logging has on biodiversity conservation in well-managed plantations and natural forests. Another approach for minimizing the effects of industrial logging on biodiversity conservation could be the promotion of sustainable forestry through forest certification. Forest certification has the potential to deliver biodiversity conservation in at least three ways (Gullison 2003).

1. It may improve the value of biodiversity in certified forests. This can be measured by the difference in the value of biodiversity in certified forests and that of forests under conventional forest management regime.
2. Certification may be sufficiently profitable for landowners to manage their forests for the production of certified timber, rather than clearing their forests for agricultural uses.
3. It may reduce logging pressure on the high conservation value forests (HCVFs) if it offers consumers the option of purchasing

**Table 9.1** Criteria and indicators of the Bhopal India Process

| *Criterion* | *Indicator* |
|---|---|
| Criterion 2: Maintenance, conservation, and enhancement; area of protected ecosystems (protected areas) of biodiversity | 2.1 Area of protected ecosystems (protected areas)<br>2.2 Number of animal and plant species<br>2.3 Number and status of threatened species<br>a. Animal b. Plant<br>2.4 Status of locally significant species—animals and plants<br>2.5 Status of species prone to over-exploitation<br>2.6 Status of non-destructive harvest of wood and non-wood forest produce |
| Criterion 3: Maintenance and enhancement of ecosystem status of natural regeneration function and vitality | 3.1 Status of natural regeneration<br>3.2 Incidence of forest fire<br>3.3 Extent of livestock grazing<br>a. Forest area open for grazing<br>b. Number of livestock grazing in forest<br>3.4 Occurrence of weeds in forests<br>a. Area b. Weed type<br>3.5 Incidence of pests and diseases |
| Criterion 5: Maintenance and enhancement of forest; growing stock of wood | 5.1 Growing stock of wood<br>5.2 Increment in volume of identified species of wood<br>5.3 Efforts towards enhancement of forest productivity<br>a. Technological inputs<br>b. Area under high-tech plantations<br>c. Area under seed production, clonal seed orchards |
| Criterion 6: Optimization of forest resources utilization | 6.1 Recorded removal of wood<br>6.2 Recorded collection of non-wood forest produce<br>6.3 Efforts towards reduction of wastage<br>6.5 Aggregate and per capita consumption of wood and non-wood forest produce<br>6.6 Direct employment in forestry and forest-based industries<br>6.7 Contribution of forests to the income of forest-dependent people<br>6.8 Demand and supply of wood and non-wood forest produce<br>6.9 Import and export of wood and non-wood forest produce |

forest products that come from a well-managed forest compared to those from a lower conservation value forest.

## 9.3 THREATS POSED TO BIODIVERSITY FROM INAPPROPRIATE FORESTRY PRACTICES

Although the causes of deforestation vary regionally (Rudel and Roper 1996; Roper 1999), industrial/irresponsible logging has been the major causal agent for forest degradation in most regions (Dudley, Jeanrenaud, and Sullivan 1995). Industrial/irresponsible logging at a very high intensity can lead to deforestation, whereas logging at a lower intensity can severely degrade the environmental value of the forest even though the forest cover may remain in the long run (van Soest 1998). Studies reveal that industrial logging could trigger deforestation by opening up vast tracts of forest to colonization, and this in turn can change the microclimate of the forest and make it more susceptible to forest fires and windthrow (Cochrane 2001).

## 9.4 CONSERVATION CONCERNS IN CERTIFICATION SYSTEMS

The 1992 United Nations Earth Summit in Rio de Janeiro defined "biodiversity" as "the variability among living organisms from all sources, including, inter alia, terrestrial, marine, other aquatic ecosystems, and the ecological complexes of which they are part: this includes diversity within species, between species and of ecosystems". This is, in fact, the closest thing to a single legally accepted definition of biodiversity. The major certification systems such as Forest Stewardship Council (FSC), Programme for the Endorsement of Forest Certification (PEFC), Canadian Standards Association, Sustainable Forestry Initiative, and Malaysian Timber Certification Council use the same, or very similar, definitions of biodiversity. Table 9.2 gives a detailed comparison of these definitions.

## 9.5 MEASURING THE IMPACT OF CERTIFICATION ON BIODIVERSITY

Although it is becoming clearer that biodiversity contributes substantially to ensuring forest ecosystem functioning and human well-being, its assessment in somewhat concrete forms remains challenging. Some of the difficulties in measuring biodiversity are the

**Table 9.2** Definition of biodiversity used in different certification schemes

| *Name of the scheme* | *Definition of biodiversity* |
|---|---|
| Forest Stewardship Council | Biological diversity is the variability among living organisms from all sources, including, inter alia, terrestrial, marine, and other aquatic ecosystems and the ecological complexes of which they are a part; this includes diversity within species, between species and of ecosystems (Convention on Biological Diversity). |
| Programme for the Endorsement of Forest Certification (PEFC) | Although not mentioned in its "terms and definitions", PEFC has based its criteria on those of the International Tropical Timber Organization |
| International Tropical Timber Organization | Biological diversity or biodiversity is not just the number of species in a particular area. Rather, it is the total variety of genetic strains, species and ecosystems that are found in nature.<br>For practical purposes, biodiversity is normally subdivided into three major hierarchical categories —variation at the genetic level within a particular species; species diversity or the number and proportion of different species in a particular area; and ecosystem diversity that describes the variation in the assemblages of species and their habitats. |
| Canadian Standards Association | Biodiversity (biological diversity)—the variability among living organisms from all sources, including, inter alia, terrestrial, marine, and other aquatic ecosystems and the ecological complexes of which they are part; this includes diversity within species, between species and of ecosystems (Environment Canada, Canadian Biodiversity Strategy). |
| Sustainable Forestry Initiative | Biological diversity, biodiversity: the variety and abundance of life forms, processes, functions, and structures of plants, animals, and other living organisms, including the relative complexity of species, communities, gene pools, and ecosystems at spatial scales that range from local to regional to global. |
| Malaysian Timber Certification Council | Biological diversity: the variability among living organisms from all sources, including, inter alia, terrestrial, marine and other aquatic ecosystems and the ecological complexes of which they are a part; this includes diversity within species, between species and of ecosystems (Convention on Biological Diversity). |

**Source** van Kuijk, Putz, and Zagt (2009)

huge variation in species, especially at lower hierarchical levels, the complex relations between species and forest ecosystem functioning, and our limited understanding of biodiversity (van Kuijk, Putz, and Zagt 2009).

A way of assessing this impact (particularly in the FSC system) is to analyse corrective action requests (CARs), that is, conditions issued when forestry operations do not fully meet certification requirements. CARs must be resolved within a specified time frame to validate the certificate. Each certified operation is audited at least once a year to ensure their continued compliance with specified standards. Through the review of publically available CARs, certification systems like FSC have the ability to demonstrate measureable performance, and a significant improvement in forest management is apparent in most cases (Mechel, Meyer-Ohlendorf, Sprang, *et al.* 2006). Listed in certification reports, the CARs thus serve as good indicators for measuring the effect of management practices that have been necessary to comply with FSC standards (caused by certification) on biodiversity. The relevant principles of Forest Stewardship Council are as follows.

- P6: Reduction of the environmental impact of logging activities and maintenance of the ecological functions and integrity of the forest.
- P9: Maintenance of HCVF—environmental values considered to be of outstanding significance or critical importance.

The impacts of reduced logging on biodiversity are discussed in Box 9.1.

**Box 9.1** Improving biodiversity through reduced impact logging

An analysis of the additive effects of reduced impact logging (RIL) on the composition and richness of tree species, macro-fauna, insects, and mammals has been made by comparing Forest Stewardship Council (FSC)-certified forests with the surrounding conventionally logged forests. This analysis showed that strict compliance with FSC requirements maintains the abundance of keystone fruit species, standing dead snags, large stems, foliage, and litter on which animals depend for food and habitat and appears to maintain plant and animal diversity relatively intact. Lowered harvest volume and minimal impacts from skid trails also help to keep the microclimate nearly intact.

*Source* Mannan, Kitayama, Lee, *et al.* (2008)

## 9.6 LIMITATIONS IN ASSESSING THE IMPACT OF CERTIFICATION ON BIODIVERSITY CONSERVATION: SOME SOLUTIONS

Despite the huge number of sustainability indicators in SFM, there is a lack of quantitative evidence of the positive impact of certification on forest biodiversity. The widespread use of conventional monitoring approaches has so far not led us to conclusively believe in the link to the effects of forest certification schemes (de Iongh and Persoon 2010). Extensive research in this direction is hindered by reasons like high costs of conventional monitoring (which normally uses the relatively small sample size) and thus does not permit representative sampling (van der Hoeven, de Iongh, Nijman, *et al.* 2000; van Kuijk, Putz, and Zagt 2009).

Researchers have, however, circumvented the need for such studies by assessing the impact of certification using the information available in public summaries of certification reports, specifically the list of CARs given to forest management units (FMUs) by evaluation teams. This has also been discussed in previous paragraphs. The CAR included in the evaluation reports demand changes or improvements in specific measures related to the certification scheme's principles and criteria. Therefore, the CAR list indicates the management actions that need to be improved. Given that FMUs have to respond to CARs and improve their management accordingly to maintain their certification, evaluation of CAR is an indirect way of measuring the impact of certification at the FMU level.

Some species of vascular plants (for example, *Mallotus* spp. and *Macaranga* spp.) have traditionally been used as indicators of forest disturbance (Kessler 1999; Slik 2001). In India, foresters have been using *Mallotus philippinensis* as an indicator of good and undisturbed sal forests. Although most evidence indicates that a single species is not a good indicator, there is some scientific evidence that single-fauna taxa (species or genera) in tropical rainforests have this predictive capacity (Azevedo-Ramos, de Carvalho, Jr., and Nasi 2002). Johns (1987) and Johns (1996) provided good examples of structural responses—in terms of changes in species composition and guild composition of avian communities—to disturbance caused by harvesting in tropical lowland forests. Several authors suggested, however, that single taxa alone are not sufficient to measure the impact of harvesting on forest biodiversity; besides they also raised serious doubts whether such an approach is helpful at all (Azevedo-Ramos, de Carvalho, Jr., and Nasi 2002; Landres, Verner, and Thomas 1988; Sheil 2001; Sheil, Nasi, and Johnson 2004). Some of them concluded that a set of indicators should be used to make accurate estimates of such impacts, instead of single indicators.

Thus two reasons for the lack of conclusive results are thought to be the high costs involved in conventional monitoring and the relatively small sample size, which does not permit quantification of forest management impacts.

To improve the situation, Sheil, Nasi, and Johnson (2004) and Sheil (2001) proposed a range of participatory approaches involving the use of local and indigenous knowledge in monitoring the impact of logging on biodiversity in affected forests. A participative, locally based approach is believed to be more cost effective as it can also cover a much larger sample area than conventional methods. But there is another school of thought that questions the wisdom of posing so much faith on locals in the absence of extensive evidence.

Danielsen, Mendoza, Tagtag, *et al.* (2007) and Danielsen, Burgess, Balmford, *et al.* (2008) also stated that in some circumstances, locally based monitoring has advantages over conventional monitoring by professionals as it can build on local capacity and relations between local people and authorities and may result in more timely management interventions.

A study conducted by the National Wildlife Federation indicates that the goal of conserving biodiversity in tropical forests using SFM principles has largely failed, but voluntary certification on private lands in temperate forests has been quite successful (NWF/NRCM/EA 2001). The degree to which certification is effective in conserving biodiversity is unknown, but it is assumed to be significant in temperate regions, though perhaps it is strongly dependent on the certifying body (NWF/NRCM/EA 2001).

Certified forest management is a complex topic and practice. Forests differ from place to place and so management practices vary. Differences in logging intensity, logging pattern and timing, the size and variety of species harvested, extraction methods, and post-harvest activities all contribute to different responses by plants and animals (Sheil, Putz, and Zagt 2010).

In the face of these difficulties, the general conclusions drawn from the literature can only be limited in nature. Similarly, a few studies (Boxes 9.2, 9.3, and 9.4) have indirectly/directly assessed the effects of certified forest management on a number of plant and animal species. They showed that populations of endangered animals increased. Due to the lack of enough literature on direct assessments of certified forest production, the available studies have concentrated on a number of good forest management practices associated with certification, such as reduced impact logging, establishment of stream-side buffers, protected areas and corridors, and HCVFs (Sheil, Putz, and Zagt 2010).

**Box 9.2** Improved fire management facilitated by forest certification

Maya Biosphere Reserve (MBR) is an area of tropical forest in Guatemala of which some areas are Forest Stewardship Council (FSC) certified. Fire incidences in the FSC-certified concessions have come down to 0.1% from 6.5% in 1998 when the first FSC certificate was issued. During the same period, fires affected 7%–20% of the rest of the reserve. The average annual deforestation rate in the FSC-certified forest concessions between 2002 and 2007 was 20 times lower than the deforestation rates within the protected areas where harvesting of wood and non-timber forest products is prohibited.

*Source* Hughell and Butterfield (2008)

**Box 9.3** Certification helps in reduced usage of pesticide in Germany

Forest Stewardship Council certification has banned the use of toxins in forests. A key issue is the precautionary treatment of timber stored in the forest with insecticides. This could be solved by the accurately timed evacuation of wood. The restraint of regeneration by graminaceous species is combated with forest management measures like reduced harvesting level in vulnerable stands. The reduced growth of grass is also less attractive for voles and, therefore, again allowed the reduction in the use of pesticides.

*Source* Hirschberger (2005)

**Box 9.4** Efficacy of sustainable forest management on biodiversity: some examples

Lagan *et al.* reported that a lowland dipterocarp forest in Malaysian Borneo (Sabah) was certified in 1997 in which reduced impact logging (RIL) and protection of high conservation value forests (HCVFs) were the measures implemented for good forest management in the study area. The study revealed that certification had a positive impact on biodiversity because the certified forest sustained denser population of endangered large animals, including orang-utans and elephants, than elsewhere in Sabah. Similarly, studies conducted in the same area on plants and soil macro-fauna showed that the use of RIL had biodiversity benefits. Diversity in tree species was equally rich in the old growth forest and in the forest harvested using RIL techniques, where climax forest and important commercial timber species of Dipterocarpaceae dominated. Moreover, the size structure of the canopy tree populations showed that dipterocarp trees regenerated well in the forest harvested by RIL, in contrast to the pioneer species that dominated the forest harvested by the conventional method.

Another study conducted by Sverdrup-Thygeson and co-workers (2008) in Norway analysed over 200 pre- and post-certification regeneration areas (mostly clear-cuts). The study revealed that in the post-certification areas, most green retention trees (left during harvesting to provide live and dead/decaying wood as habitat) were present and the buffer zones along rivers, bogs, and lakes were wider than those in pre-certification areas, which are expected to benefit biodiversity (but these benefits were not measured directly). However, 21% of the post-certification areas did not have sufficient retention trees to comply with the certification standard. Concerning the management of small swamp forests and the damage done due to off-road transport, little improvement was observed. Table A summarizes the direct effects of certification on different taxa.

**Table A** Summary of the direct effects of certification on different taxa

| *Certification* | *Comparison of certified forest landscape with reference managed forest landscape* | *Comparison of certified forest landscape with undisturbed forest landscape* | *Tropical* | *Temperate Boreal* |
|---|---|---|---|---|
| Forest biodiversity | +? | ND | | × |
| Mammals | + | | × | |
| Trees | + | = | × | |

*Contd...*

**Box 9.4** *Contd...*

**Table A** *Contd...*

| *Certification* | *Comparison of certified forest landscape with reference managed forest landscape* | *Comparison of certified forest landscape with undisturbed forest landscape* | *Tropical* | *Temperate Boreal* |
|---|---|---|---|---|
| Soil macro-fauna | +? | | | |
| Landscape features | = | | | × |

ND – not determined

**Note** +(=.-) means that the application of certification or a good management practice resulted in a positive (insignificant; negative) effect on biodiversity compared with the reference.

**Source** van Kuijk, Putz, and Zagt (2009)

Despite the apparent differences in the rigour with which biodiversity concerns are addressed under different certification systems, the planning, supervision, and basic good management practices required by all of them serve to mitigate many of the harmful environmental impacts of logging and other forest management activities. The review also shows that the impacts of certified forest management on biodiversity cannot be assessed without a clear idea of the relative importance of species and management objectives (van Kuijk, Putz, and Zagt 2009). Similarly, these studies confirm that despite their better performance, certified forests are not fully equivalent to undisturbed or primary forests in terms of biodiversity.

In general, some of the studies reviewed suggest that forest certification has positive biodiversity benefits. This is in agreement with information gathered from general discussions with certifiers and forest managers with experiences in the field (Sheil, Putz, and Zagt 2010).

## 9.7 IMPACT OF CERTIFICATION ON PLANTATION MANAGEMENT

A 2009 analysis of FSC-certified plantations and natural/mixed forest operations worldwide, audited by Rainforest Alliance/SmartWood Programme, reveals that on an average, certified forestry operations designate 16% of their total forest area as strict reserves. Operations with a higher percentage of plantations have significantly larger strict reserve areas. This trend may be due to FSC principle 10.5, which requires that a proportion of plantation forest area be restored to natural forest cover. Operations with a relatively high amount of plantation forest area had higher reserve area than those with low or no forest plantations. The relatively high amount of strict reserve area found on plantation forests may be seen as a sign that forest companies are implementing and auditors are evaluating criterion 10.5 in a meaningful way.

Box 9.5 discusses the role of forest certification on biodiversity conservation.

**Box 9.5** Role of forest certification on biodiversity conservation: is there demonstrable improvement?

Over the years, the belief has grown that responsible management of forests for timber production may also make a positive contribution to biodiversity conservation. The promotion of socially and ecologically sound forest management through forest certification is now widely embraced as a strategy to conserve the world's forests and the biodiversity they contain. Increasing the extent of certification in the tropics remains a goal for many organizations, including some international conservation non-governmental organizations. So far, so good; but many details about the effectiveness of certification remain uncertain, including those on biodiversity.

Regarding biodiversity conservation in certified forests, many authors involved in certification and/or conservation of tropical forest biodiversity have given their views on whether certification is a good conservation strategy for tropical forests. Their articles reported on practical experiences from concessions and community forests, challenges of monitoring biodiversity, high conservation value forests, and a range of other subjects. The results of a dedicated on-line survey provide additional context.

A positive general impression emerges, but not without qualifications. Most authors and respondents agree that certification has helped to improve management practices and conserve forest biodiversity within certified forests in the tropics. However, the true extent of conservation benefits remains unknown due to a lack of rigorous and independent information. Many researchers have agreed that certification is not equivalent to full conservation and point at the limitations of certification in reducing deforestation rates.

*Source* Sheil, Putz, and Zagt (2010)

## 9.8 CONCLUSIONS

The challenge for forest plantation managers, certifiers, and biodiversity conservationists and researchers is to develop and promote appropriate forest certification from a credible proposition to a demonstrated asset in the suite of instruments available for forest biodiversity conservation.

To further inform the trade-offs between biodiversity and the social and economic interests of forest management accepted by certifiers, certification schemes will have to provide quantitative, field-based evidence of species' responses to forest management practices and propose modifications if required.

# 10
CHAPTER

# Different Certification Systems: Convergence Issues

## 10.1 ISSUE OF HARMONIZATION AND POINT OF DIVERGENCE

Of late, forest managers/owners are confronted by a plethora of new standards of certification or verification schemes for different purposes. For example, certification schemes exist in the consideration of biomass, energy, carbon sequestration, organic, labour, fair trade, and legality. Today, the market seems to be expecting some sort of mutual recognition and harmonization of the major international standards, if not a common global certification standard (Fischer, Aguilar, Jawahar, *et al.* 2005). The major drawbacks of proliferating certification systems/programmes are given in Box 10.1.

This developing proliferation of standards has added a lot of confusion and higher costs to both producers and consumers. It also

**Box 10.1** Drawbacks of proliferation of forest certification programmes

- It creates confusion in the minds of certified forest producers and owners.
- It creates confusion in the minds of the purchaser of certified forest products.
- It leads to broad inefficiency in forest and market.
- If not controlled, in extreme cases, it may theoretically lead to the defeat of the very basic purpose of certification, that is, no incentive for sustainable forest management.
- It adds cost to certified forest product producers to conform to different norms for different certification programmes/systems for different considerations. At the same time, it is a costly proposition for purchasers also.

entails the risk of unequal requirements applying to different systems. Hence, the certification systems urgently require harmonization and coherence if they have to achieve their intended credibility, efficiency, effectiveness, and scope.

The market expects that processes and standards would evolve and improve over time, and the disparate approaches currently used among forest certifying organizations are likely to coalesce into a common framework, minimizing confusion in the minds of both producers and consumers.

The level of public discord between the two leading certification frameworks has been particularly intense in the USA (Box 10.2). A legal action filed in 2008 was associated with the distribution of settlement monies under the Softwood Lumber Agreement between the USA and Canada. The issue in question was the award of some monies to one certification programme without equal consideration of what were argued to be similar benefits or services offered by another. This action included Forest Stewardship Council (FSC)-USA as a plaintiff. In 2009, Forest Ethics filed a complaint about abuse of non-profit regulations by the Sustainable Forestry Initiative (SFI)

**Box 10.2** Divergence in certification systems: ramifications

In October 2009, the Forest Stewardship Council (FSC) and the Programme for the Endorsement of Forest Certification (PEFC) got engaged in a public dispute following a publication of an FSC report titled "Comparative analysis between the Forest Stewardship Council controlled wood requirements and Programme for the Endorsement of Forest Certification, Programme for the Endorsement of Forest Certification Germany and Sustainable Forestry Initiative". The report suggested that certification of systems endorsed by PEFC is insufficient evidence of conformity to the FSC-controlled wood standard and, therefore, cannot be mixed with FSC-certified material without further examination. In response, the PEFC International described it as an attempt to undermine alternative approaches to certification, "disregarding" years of work to build an international consensus among forest certification schemes. The FSC report also had immediate practical implications because numerous wood traders supplying FSC-labelled products had relied heavily on PEFC-certified wood to demonstrate conformity to FSC-controlled wood standard. These traders now have no option but to introduce additional potentially costly controlled wood standard risk assessment and verification procedures.

*Source* Oliver (2010)

and misleading eco-label claims. These claims were filed before the Internal Revenue Service and the Federal Trade Commission. Within weeks of the Forest Ethics' filing, forest-sector companies responded with a complaint filed against FSC and the US Green Building Council (USGBC) for misleading eco-label claims of their own and for exclusionary practice against equivalent certification programmes. Whether any (or all) of these legal actions have any merit is to be debated in public forums, but such activity may cause increased public and decision-maker cynicism and dampen enthusiasm for forest certification. Hence, advocates of forest certification may be advised to take note of the words of the US National Association of State Foresters in their October 2008 policy statement on forest certification (Oliver 2010) (Box 10.2).

While in different ways certification schemes like the American Tree Farm System, FSC, and SFI include the fundamental elements of credibility and make positive contributions to forest sustainability, the proponents of individual certification programmes often promote their option as the best or the only option. Their claims have little to do with quality and everything to do with marketing and selling their programme. No certification programme can credibly claim to be the "best", and no certification programme that promotes itself as the only certification option can maintain credibility. Forest ecosystems are complex, and thus a simplistic one-size-fits-all approach to certification cannot address all sustainability needs. Thus harmonization is the need of the hour.

While these standards attempt to address the operations of the entire industry, they often build niche markets for specialty products (for example, organic produce and herbal produce). The programmes are the result of the combined efforts of non-governmental organizations, corporations, and other stakeholders, often with government support (Atyi and Simula 2002). The FSC scheme established in 1993 was the early front runner, and for a number of years remained the sole international accreditation organization. Its decision-making structures were innovative in the way they brought together divergent interests and, through its 10 principles for forest stewardship, helped define what sustainable forest management (SFM) should be. Other certification schemes/programmes followed, initially at the national level (for example, the Canadian Standards Association scheme and SFI in the USA). However, the main challenge to FSC came from a regional initiative launched in 1999—the Pan-European Forest Certification Council (PEFCC). The success of PEFCC can be attributed to its working with European industry interests and small forest owners that led to its relaunch in 2003 as the Programme for the Endorsement of Forest Certification (PEFC). The largest area of certified forests worldwide comes under the PEFC brand (Nussbaum and Simula 2004). However, the emergence of the diversity of national, regional, and global schemes is causing great confusion among consumers and hindering competition among suppliers, who may not be able to afford multiple certifications for multiple clients.

## 10.2 CONFLICTS AND/OR INCOMPATIBILITY BETWEEN LEGAL SETTINGS AND CERTIFICATION STANDARDS

In some cases, there could be conflicts between national laws and forest certification standards. For example, the ownership of land, usufruct rights of forest products and services, and the sharing of responsibilities between the government, local people, and forest concessionaire may be locally defined in a way, which may not conform to the requirements of voluntary certification. Thus it may be considered "illegal" based on public law or unacceptable for voluntary certification standards, but it may still be in accordance with customary laws and the traditional rights of local people living in and around forests (Atyi and Simula 2002).

The above points are just a few examples of conflicts and disputes, which may need to be addressed before forest certification can work effectively in practice. Locally developed standards through an inclusive and participatory process within an appropriate international framework may result in performance requirements relevant to and achievable by stakeholders, communities, and other forest owners in specific country conditions. This process of standard development may reduce the possibility of the above-mentioned disputes occurring.

The conflict between certification systems may also add to an element of uncertainty. The competition war for market share continues between the various major forest certification schemes, including FSC and PEFC systems. However, recent events suggest that the potential for compromise, constructive engagement, and mutual recognition between the two major systems seems as far away as ever (Oliver 2010).

## 10.3 SOME POSITIVE DEVELOPMENT TOWARDS HARMONIZATION

But there are some bright examples of harmonization like the Principles, Criteria, and Indicators (PCI) of the African Timber Organization (ATO), which have been harmonized with the International Tropical Timber Organization's criteria and indicators, including sub-indicators directly applicable to certification audits.

Another approach has been adopted by FSC, which has developed its own principles and criteria for well-managed forests independently of any other international sets through a consultative process. The international/regional sets of criteria and indicators share a number of common elements, and they also largely cover nine of the FSC principles.

Despite these examples of a shift towards compatibility, it needs to be emphasized that criteria and indicators for SFM are neutral assessment tools for monitoring trends and, therefore, cannot be used as substitutes for minimum agreed-upon forest management standards, which form the core of forest certification (Atyi and Simula 2002).

## 10.4 CONVERGENCE BETWEEN CERTIFICATION SYSTEMS

### 10.4.1 Structural Issues

There are at least five possible avenues for the harmonization of certification in the forestry sector: (1) international standards; (2) recognition of equivalence (for example, through mutual recognition between existing systems); (3) regional or international cooperation; (4) the bottom-up harmonization of standards; and (5) unilateral recognition (ITTO 2008).

Over the years, many of the issues that previously divided the certification systems have become much less distinct. The largest certification systems now generally have the same structural programmatic requirements (Table 10.1). This convergence is partially driven by the need for various systems to conform to equivalent international standards, either for mutual recognition under a

**Table 10.1** Characteristics of major forest certification programmes (as in 2010)

| *Programme* | *Third-party auditor* | *Chain of custody* | *Public reporting* | *Stakeholder consultation* | *Independent governance* | *On-product* label |
|---|---|---|---|---|---|---|
| ATFS | Yes | Yes | Yes | Yes | Yes | No |
| CSA | Yes | Yes | Yes | Yes | Yes | Yes |
| FSC | Yes | Yes | Yes | Yes | Yes | Yes |
| PEFC | Yes | Yes | Yes | Yes | Yes | Yes |
| SFI | Yes | Yes | Yes | Yes | Yes | Yes |

ATFS – American Tree Farm System; CSA – Canadian Standards Association; FSC – Forest Stewardship Council; PEFC – Programme for the Endorsement of Forest Certification; SFI – Sustainable Forestry Initiative

**Note** CSA has adopted the PEFC on-product label and discontinued using the CSA on-product label.

**Source** Fernholz, Howe, Bratkovich, *et al.* (2010)

framework like PEFC or to ensure acceptance in various public and private sector procurement policies (Oliver 2010).

With the changes made since 2004, many of the previous differences between forest certification programmes are getting much less distinct. Table 10.1 shows that each programme generally has the same structural programmatic requirements, although the required content and level of detail provided by each may vary considerably. For example, each programme has elements of stakeholder consultation and tribal people's rights, but this apparent similarity highlights remaining distinctions (Fernholz, Howe, Bratkovich, *et al.* 2010). This is elaborated by the fact that in the FSC programme, consultation with indigenous people is required for certification of all forest management operations, while the SFI programme requires consultation with tribal people only when the certification assessment is for public forest lands. These are similar common considerations, but they get highlighted in a particular programme to make a unique marketing proportion.

Box 10.3 discusses some differences found within the programmes and their respective standards for evaluating forest management practices.

## 10.5 CONVERGENCE IS HAPPENING

The process of convergence is continuing. For example, PEFC has started engaging stakeholders to develop a new standard, setting out international requirements for SFM. For schemes like PEFC, this

**Box 10.3** Issue of genetically modified organisms: the point of distinction

Several additional differences can be found within the programmes and their respective standards for evaluating forest management practices. The most prominent distinction that remains between the Sustainable Forestry Initiative (SFI) and the Forest Stewardship Council (FSC) is that FSC has a comprehensive ban on the use of genetically modified organisms (GMOs). This restriction in FSC is included in the FSC certification standard (FSC Criterion 6.8) as well as within the FSC Controlled Wood Standard (FSC-STD-40-005 V201), meaning that FSC-labelled products cannot contain GMOs or come from a forest where GMOs are planted. On the other hand, SFI does not have a prohibition regarding GMOs. Given that GMOs are not included in commercial forest product production in North America, the impact of this distinction is currently more philosophical than physical.

*Source* Fernholz, Howe, Bratkovich, *et al.* (2010)

is a departure from its earlier concept in which prominence was given to national forest certification standards developed in line with international sustainable forestry principles agreed through intergovernmental, multi-stakeholder processes (Oliver 2010). The PEFC standard also tries to establish the principle that there should be no infringement of "legal, customary, and traditional rights" without free and informed consent. Thus some remaining key distinctions between PEFC and FSC have been removed.

## 10.6 SCOPE FOR AN OMNIBUS CERTIFICATION SCHEME

As observed earlier, over the last few years a range of different certification schemes have been developed. This has led to considerable curiosity in the market about which schemes are more credible, visible in the market, and economical. The issues of mutual recognition between these numerous schemes and of minimum key attributes to be an acceptable scheme are some important questions always in the mind of the producer and purchaser. The effort to reconcile this debate has not always been very harmonious as there have been considerable differences of opinion between the supporters of different schemes (Handford and Nussbaum 2004). This is justified by the fact that specific certification schemes offer a "distinct certified forest product", which is liked by the targeted segment of purchaser.

Though it is possible to develop a certification scheme that delivers multiple objectives, it would have great complexity and be much more costly. Hence, the normal tendency is to develop a scheme that delivers one set of simple and economic operating objectives (Handford and Nussbaum 2004). But it is the market that has a great role in evolution of any such omnibus scheme catering to different objectives and making a trade-off between complexity and economics. Box 10.4 expounds on the holistic certification of forests.

**Box 10.4** Quest for holistic certification of forests

The wide application of certification of forests still remains problematic because of costs associated with its implementation and ongoing compliance. The cost issue gets compounded by the multiplicity of certification. Internationally, some attempts are being made to deal with this through initiatives such as group certification and more recent projects such as the Fairtrade–Forest Stewardship Council (FSC) dual certification pilot.

In a similar vein, FSC's certification agency, Accreditation Services International, is working to ensure full conformance with ISO 17011, the international accreditation standard, by 2011. This would remove another difference with the Programme for the Endorsement of Forest Certification (PEFC), which requires that accreditation bodies conform to the International Organization for Standardization (ISO) rules (Oliver 2010).

Hence there is still a long path to travel to reach holistic certification. Another issue is that the commodity focus of most certification programmes (here mainly wood and agricultural commodities) can lead to a fragmented view of development opportunities rather than a holistic view of value creation.

Assisting various communities to find the most value-enhancing outcome for them in conformity with different certification opportunities is a big challenge.

*Source* <www.forestry.about.com/od/environmentalissues/a/sustain_forests.htm>

Some additional efforts are being taken in the direction of harmonization. For example, entirely new systems of certification are being designed for sustainable biofuel production and for carbon sequestration, which overlap with (but at the same time may have the potential to come into conflict with) existing systems of SFM certification. To ensure rational decision-making with respect to appropriate land use and forest management objectives and to reduce costs imposed on the forest sector, there is a growing need to harmonize various forest certification frameworks being developed for sustainable timber production, sustainable biomass production, and carbon sequestration (Oliver 2010). But to best enhance the role of forests in climate change mitigation, the various standards and certification systems that emerge from this process would require to avoid creating perverse incentives, for example, encouraging deforestation. It should also promote a "cascaded" use of wood (Oliver 2010).

## 10.7 REFORMS NEEDED

Layton, Guynn, and Guynn (2002) reviewed the metrics used and approaches taken to certification (specifically dealing with wildlife and biodiversity evaluations). They concluded that auditors often use professional judgment as the basis for determining if a criterion or indicator is met. Because audit teams vary from owner to owner, the consistency of evaluation can be questionable. Thus certification bodies will have to struggle hard to meet with objective measures of performance, especially when evaluating different indicators, which is yet another important issue.

As the use of genetically modified organisms is expected to expand, FSC and other programmes may face challenges with enforcing this standard, and it may prove to be a defining difference between programmes. One of the unique dilemmas that may develop could be the potential use of genetic modification technologies for species restoration efforts (Box 10.5).

There is an effort being made for achieving some degree of compatibility between FSC and PEFC accreditation options. The fundamental objectives and goals of both FSC and PEFC are similar but achieved in different ways. However, as mentioned earlier, these are essentially two competing timber and wood products marketing schemes, and their relative success will depend as much on the success with which they are able to assure the market of product standards and assure the buyer groups about the fundamental quality of the principles and criteria (Eyre and Mundy 1999).

The choice of programmes for certification is explained in Box 10.5.

**Box 10.5** Choice of programmes for certification, given their proliferation

Even though the distinctions between forest certification programmes may have become more and more subtle, deciding which programme to support or participate in is still based on specific values and circumstances. Furthermore, choosing to certify or not to certify is a decision still to be made by many forest owners and managers. The blurring of the lines does not mean there are no real reasons why organizations continue to choose one programme over another.

As mentioned in a 2004 report, "for organizations weighing the choice between third-party certification systems, careful consideration must be given to the perceived bias in each. For example to some, SFI certification is like having the fox guarding the "hen house". Alternately, submitting to Forest Stewardship Council oversight can be viewed as giving in to the added control and costs demanded by tree-huggers. Often, resistance to certification by either system is a result of fear, based on hidden, or subconscious, bias".

It is not a good organizational policy to make decisions based on fear and bias. Organizational stakeholders deserve better, and forest managers need to recognize this and develop a comparison based on analysis of the actual merits of the varying standards instead. Increasingly, the final decision regarding certification participation has to be based on the perceived potential advantages and merits of the marketplace.

*Source* Fernholz, Howe, Bratkovich, *et al.* (2010)

## 10.8 CONCLUSIONS

Convergence between certification systems may contribute to their consistent treatment in different policies and evaluations of acceptability. Seeking recognition for a national forest certification scheme on an individual basis under public timber procurement policies has so far not proved easy. The bottom-up harmonization of standards might facilitate exporters by offering the flexibility to supply timber under different international labels depending on demand. It would also allow the commencement of the national standard development process in the absence of a decision on which international system to apply during implementation.

For better or worse, at the end of the day, all forest certification programmes have more or less the same goal and the same measure of success. Each programme is trying to identify wood as a responsible, accountable, and assured green product in the marketplace. There is unhealthy competition because each certification programme has to measure its success by the volume (and value) of products that are being bought and sold in the marketplace under their respective labels. In fact, every certification programme would like to be the "forest products company". They are not sensitive to the number of hectares certified or the number of certificates issued even if marketplace transactions are not occurring. Without a functioning marketplace, which includes robust supply and demand, there is little legitimacy to an argument that any of the certification programmes has the capacity to provide a lasting impact on forest resources. Here harmonization of different schemes of forest certification can play an important role in restructuring and achieving long-lasting impact.

# 11

CHAPTER

# SFM, Forest Certification, and Climate Change Regulations

## 11.1 CLIMATE CHANGE IMPERATIVES

The forest sector is undergoing a massive metamorphosis and is trying hard to respond to diverse and often apparently conflicting stimuli. On the one hand, it is facing a sustained increase in demand for its products and services. On the other hand, the sector is trying to evolve sophisticated management and compliance tools to secure public goods, particularly those associated with the environment such as carbon storage (Eyre and Mundy 1999). Thus the role of forest in climate amelioration through mitigative and adaptive ways is assuming importance in the context of the Kyoto Protocol and other related voluntary efforts for ensuring greenhouse gas (GHG) mitigation.

Current forest certification initiatives are developing against a rapidly changing background of international and national initiatives in forestry, biodiversity, conservation, environmental management systems, and trade, many of which also aim to achieve sustainable forest management (SFM) regimes on the ground. A sustainably managed forest is expected to be more resilient to climate change and to have the ability to act as a barrier to some of the impacts of the most serious climate change impacts (such as flooding, storms, and pests) thereby helping to protect many vulnerable populations. The subsequent sections explain how certification and SFM can be important tools for ensuring climate change mitigation. Certification promotes and assures SFM, taking explicit account of environmental, economic, social, and cultural dimensions of forest management, conservation, and development in a holistic manner.

## 11.2 SFM AND CLIMATE CHANGE

The effect of climate change and the Kyoto Protocol is very significant for the forestry sector's evolution and development as it may facilitate or act as a catalyst to capture the economic impact of different environmental values from the forest. This development is also resulting in the design of more effective management options and regimes for the forest.

Attempts on these lines have been made through the Kyoto Protocol and similar other mechanisms (many of them voluntary in nature) for climate regulations at international and regional levels attempting (directly and indirectly) to permanently change the fundamental relationship between products and services normally associated with forestry. On similar lines, SFM systems are being introduced to establish an environmental management standards and chain of custody for forest products. This is being done to satisfy the consumer demand for sustainably produced forest products and to meet the increasingly environmentally rigorous national regulations for forest. However, this shift towards SFM has happened quite independently of the Kyoto Protocol or similar international efforts to reduce GHGs. It appears that there is a very strong connection between the two major drivers of change. Their connection is quite evident in terms of the strong similarity of management regimes for SFM and carbon sequestration forestry projects (Eyre and Mundy 1999). A sustainably managed production forest is carbon neutral in the long term, that is, there is no long-term net emission or sequestration of carbon (ITTO 2011).

Forestry projects, as developed under the Kyoto Protocol rules, require a new level of economic and project security if additional revenue streams are to be realized from the sale of carbon offsets. On the other hand, increased responsible management requirements of SFM-type certification systems reduce risks associated with forestry operations, although they are unable to shield projects against some risks that could threaten the economic viability of the project (Eyre and Mundy 1999). Hence, when carbon offsets from forestry projects are being sold, and particularly where these are sold forward under one of the flexible mechanisms, the benefits of risk transfer and insurance can be of considerable interest to both buyers and sellers of these offsets if the forest is managed under SFM regime and also certified.

## 11.3 FOREST CERTIFICATION AND CLIMATE CHANGE

Global concern for climate change and increased interest in the role of forests in mitigation strategies have major implications for the practice of forest certification as well. While forest certification systems such as Forest Stewardship Council (FSC) and Programme for the Endorsement of Forest Certification (PEFC) were developed with the major aim of supplying certified sustainable timber products to market, climate change has significantly broadened the scope of such economic products to include wood for energy production, non-timber forest products, as well as carbon sequestration for climate mitigation (Oliver 2010).

Thus new opportunities are arising from the recognition of certified sustainable forest products and services in meeting the requirements for carbon offset through projects under compliance or voluntary mechanism. This requirement of certified products and services (including carbon) will remain in the future as well. For example, in a programme like the Reduced Emissions from Deforestation and Forest Degradation (REDD), which will be soon integrated into an international agreement and might replace the Kyoto Protocol when it expires in 2012 (Oliver 2010), the requirement of SFM and certification would remain. The aim of REDD+ is to provide financial incentives to help tropical countries voluntarily reduce national deforestation, conserve and sustainably manage their permanent forest estates, and increase forest cover through reforestation and afforestation.

Thus REDD+ could simultaneously mitigate climate change (through carbon capture and storage), conserve biodiversity, protect other ecosystem goods and services, increase income for forest owners and managers, and help address issues of forest governance (ITTO 2011).

This background promises to offer a new and significant source of financing for certified sustainable forestry operations. To fully exploit these new opportunities, SFM and forest certification systems may need adaptation, like including explicit recognition of the need to monitor and increase carbon stocks over time (Oliver 2010).

Climate change issues assume even greater significance in the role of forest certification as a mechanism to counter preservationist attempts or to simply lock up forests as a carbon store, which in many cases is likely to be a suboptimal solution. On the other hand, it may also ensure increased market access for various sustainably sourced forest products (Oliver 2010). The benefits of climate change mitigation by substituting more fossil-fuel-intensive products with such products—particularly in the construction sector—can be considerable. This could still be an important factor in the future for the spread and popularity of SFM and forest certification.

## 11.4 LINKAGES EXPECTED BETWEEN VARIOUS CLIMATE CHANGE MECHANISMS THROUGH RISK MANAGEMENT

SFM, forest projects under the Kyoto Protocol, and forest certification appear to have some similarity from the point of view of risk management. Attracting finance, running SFM operations, getting certified, and developing secure carbon credits require compliance with some criteria and addressing related risk issues in a cost-effective manner. The logical response to all the above is to design a good project, which can enhance the effect by appropriate risk transfer (Eyre and Mundy 1999). However, the latter is dependent on the former as risk transfer is never a substitute for good management.

This correlation between sound management, risk transfer, and financially viable operations reinforces the view that adoption of SFM and getting certified are essential and so is establishing sound carbon-offset projects in the forestry sector (Eyre and Mundy 1999).

In this context, risk management and the insurance industry may perform a linking or bridging function between political processes and the market (Eyre and Mundy 1999). This is because of the fact that the attainment of SFM, forest certification, and multi-use forestry (including carbon sequestration purpose) is largely a political process. This intends to establish a supportive regulatory framework, which has government support in the market.

The efficiency of Kyoto Protocol or voluntary mechanisms is currently based on the resolution of outstanding political and consequent policy issues. By providing a value to one indirect use value (that is, carbon sequestration), it also provides a vicarious value to other SFM products, which would have otherwise remained uncaptured (Eyre and Mundy 1999).

Consequently, the financial flows attributed to carbon-offset projects can also be a mechanism to transfer financial flows into SFM, which in turn promotes forest certification. Given the declining funding support in the forestry sector, the apparent opportunity costs for the private sector would lead to a greater support for SFM over unsustainable forestry. This will be possible as these extra flows from carbon benefits could meet financial shortfalls.

There are various ways of transferring risk, and it is important for forest owners to choose the most cost-effective one for their needs. Given the need for carbon credits, which are secure and transferable notwithstanding their initial source, asset fungibility and security will be critical in ensuring that the forest sector plays its full role in the implementation of Kyoto Protocol or even under voluntary regional regulatory mechanisms.

## 11.5 SIMILARITY BETWEEN SFM AND CARBON FORESTRY CREDIT PROJECTS: MECHANICS OF REGISTRATION AND ACCREDITATION

There are functional linkages between accreditation of forest resources under SFM and registration under carbon credit projects (whether under the Kyoto protocol or not), or any voluntary mechanism. Both schemes require accreditation and certification procedures, although mechanisms for forestry carbon projects are quite cumbersome (Eyre and Mundy 1999)

- Accreditation under SFM involves compliance to a series of criteria covering institutional frameworks, socio-economic issues, conservation considerations, and carbon cycles. All these compliance obligations may be broadly similar to those required under the Kyoto Protocol or any voluntary mechanism (where the project will have to follow the relevant principles, rules, and guidelines in detail, including verification and reporting). Under clean development mechanism (CDM), the carbon forestry project needs to conform to the principles of sustainable development also (which may or may not be the case in voluntary mechanism).
- SFM also covers general environmental compliance criteria such as biological diversity, forest health, and the environmental protective functions of forests. All these are in compliance with global

environmental protocols and their objectives, which were also the foundation stones of the Kyoto Protocol and other voluntary mechanisms for controlling GHGs.

Thus SFM and global carbon-reducing mechanisms (regulatory, like Kyoto Protocol, and many voluntary ones) are also likely to have the greatest degree of mechanistic compatibility in terms of the accreditation system covering the measurement and monitoring of productive functions and project output. SFM provides a toolkit for generating operational and management data, guidelines on procedures, and operational restraints and objectives, all of which will be directly beneficial if the same forest project attempts to register for carbon credits either under regulatory (including CDM) or any voluntary mechanism.

In addition, a key benefit of accreditation of a CDM project under SFM is that it ensures optimization of management procedures and risk mitigation, resulting in increasing confidence in management capacity and, therefore, probably enhancing both the value and liquidity of any carbon offsets generated.

## 11.6 DIVERGENCE IN OBJECTIVES AND GOALS OF CARBON REGULATORY MECHANISM, SFM, AND FOREST CERTIFICATION

Although the objectives of carbon regulatory mechanisms (Kyoto Protocol included) and SFM, forest certification is set to create a more sustainable world. The scope and impact of a carbon regulatory mechanism are very different from SFM as it is about increasing and protecting carbon stocks as well as reducing anthropogenic carbon emissions. For example, the Kyoto Protocol's ultimate focus is on the protection and expansion of carbon stocks (Eyre and Mundy 1999), whereas the objective of SFM is to improve the management of forests for meeting social, economic, cultural, and spiritual needs of current and future generations. Also SFM is limited to the management of carbon stocks in forests, but it is not necessarily or directly linked to the increase in those carbon stocks.

Therefore, while there is a high degree of congruence between the environmental objectives in the two processes, the operational requirements may be mutually exclusive (Eyre and Mundy 1999). For example, the establishment of monoculture plantations for carbon sequestration (if eligible) may be desirable for carbon regulatory schemes, but it may not meet biodiversity conservation criteria and thus is undesirable for SFM.

As yet, there is no formal linkage between carbon regulatory mechanism like the Kyoto Protocol and SFM. Under the Kyoto Protocol, there are specified obligations for any participating party. The management compliance under SFM requires operational conditions and targets that improve environmental performance as well as operational control and knowledge. These are also similar to key requirements for any carbon sequestration project registered under the Kyoto Protocol (through either CDM or any voluntary mechanism) (Eyre and Mundy 1999). Therefore, it is evident that a forestry project certified under SFM would be at a great advantage if it wishes to participate under the Kyoto Protocol or any other regulatory/voluntary mechanism.

The fundamental output of SFM is normally conventional forest products that can utilize and gain access to premium markets. Thus the commercial benefit of SFM and subsequent forest certification is in the selling of conventional carbon stocks (mainly timber) for further processing or conversion, generating instant income. The Kyoto Protocol or other regulatory/voluntary mechanisms, however, seek to create a new product, the "carbon offset". But to qualify for this, "carbon stocks" must expand (in CDM, over a "business as usual" baseline) and remain as carbon for a very significant time (Eyre and Mundy 1999). Therefore, this leads to a fundamental contradiction between the two systems, as only carbon offsets can be achieved in one, while conventional outputs remain as carbon stocks in the other. The comparative income streams and costs associated with accessing these streams will also be different.

## 11.7 COMMON FUTURE OF CARBON REGULATORY MECHANISM (KYOTO PROTOCOL), CERTIFICATION OF FORESTS, AND SFM

It is quite likely that in the future there will ultimately be a Kyoto Protocol-like global commitment and mechanisms whereby nations and regions agree to or voluntarily resort to a gradual reduction in less emission of GHGs. At the same time, SFM would be practised as a voluntary (or even regulatory) form for commercial reasons. As a voluntary scheme, participants would like to practise SFM to extract a commercial advantage out of the market or to secure their position within the market. It will become obligatory only if there is a global commitment like the Kyoto protocol, which forces forest product markets to exclude all outputs from non-sustainable forests (Eyre and Mundy 1999). While non-governmental organizations have been quite successful in using the power of the market to expand the concept of forest certification, there is a long way to go before there is general global commitment and it becomes mandatory.

Of late, forest certification is also considered an important initiative to motivate industrial companies producing and releasing carbon dioxide into the atmosphere to balance their industrial outputs of carbon by sequestering an equivalent amount of carbon in forest biomass. Thus forest certification can be a tool to ensure that these goals could be satisfied at the ground level in the most efficient and effective manner (Vogt, Larson, and Gordon, *et al.* 2000).

Thus it can be expected that SFM and forest certification would be practised as long as there are commercial advantages. This also means that this trend would continue as long as there is a premium on forest outputs, as long as it ensures market access, or it brings other benefits associated with management in terms of access to resources or finance (Eyre and Mundy 1999). Besides this, forest certification assures the purchaser of carbon offset from this forest for various quality considerations.

## 11.8 CONCLUSIONS

Thus the forest sector is entering a phase of profound policy and operational change, in response to which it will have to develop new approaches to policy and operational risk management. If it responds to this challenge by adopting innovative financial and insurance instruments, it can reduce the risks and environmental costs associated with the change towards SFM and multi-use forestry (including carbon forestry). Also this trend can bring down the incremental costs of internalizing environmental externalities, the greatest of which is arguably climate change. Thus the development of SFM and forest certification would be desirable for an effective and efficient carbon regulatory regime be it a Kyoto Protocol-like regime or other voluntary mechanisms. All these three concepts are supplementary and complementary to each other.

# 12 CHAPTER

# Developing Certification of Forests in India: Rationale and Options

Certification promotes and assures sustainable forest management (SFM) taking explicit account of environmental, economic, social, and cultural dimensions of forest management, conservation, and development in a holistic manner. Certification of forests is increasingly being used by a number of governments in both timber-producing and timber-importing countries as a means of identifying and encouraging good forestry practice. All over the globe, governments have promoted or have been requiring certification by making it a condition of awarding concessions or resource utilization rights. For example, in case of the Mayan Biosphere Reserve in Guatemala, concessions were awarded to the community in lieu of the forest being certified (Soza 2003). Similarly, the consuming countries are increasingly requiring their suppliers of wood products to ensure that their sources are legal and sustainable. Certification is being used as a measure of sustainability. Governments are also putting pressure on suppliers and forest managers to achieve certification of forests and products (Higman, Mayers, Bass, *et al.* 2006). Although there may be little choice about which certification scheme is acceptable to governments, there is scope to be flexible about the timescale for implementation. For example, government procurement requirements being developed in the UK have two distinct categories of suppliers. This starts with establishing the legality of the forest source and promotes sourcing of forest products from legal sustainable sources, which may be demonstrated by certification (Higman, Mayers, Bass, *et al.* 2006).

Box 12.1 discusses how management goals can be met with the help of forest certification.

**Box 12.1** Meeting management goals with the help of forest certification

Certification can be helpful in devising a framework for meeting internal management goals within an organization. The other points that need to be taken care of before choosing a particular certification scheme are as follows.

If any of the drivers for certification (market, investor demand, or customer demand) are likely to become more important in the future, it might be worth choosing a certification scheme that will satisfy those needs at the same time. It may be possible to choose certification against one scheme while still ensuring compatibility with another scheme so that the elements of a second scheme can be added at a later stage if it becomes desirable.

*Source* Higman, Mayers, Bass, *et al.* (2006)

## 12.1 BASICS OF CERTIFICATION OF FORESTS IN INDIA

Certification of forests is not a new concept for the global forest sector, which is already exposed to some or other form of certification of products (if not services). But forest certification is yet to emerge and evolve in India in a big way.

The initial attempts to attain certification in any country are often difficult. This holds good for a country like India as well. The situation calls for all parties interested in certification to work together from the start to ensure that everybody or rather every stakeholder has adequate information about the concept of certification and required certification processes to be followed. Forest managers/owners and related stakeholders should visualize benefits and all consequences from the certification (both direct and indirect).

Since India has diverse agro-climatic regions and diversity in vegetation, it has very distinct socio-economic contexts. This makes it very complex for the forestry sector, which needs to satisfy basic subsistence needs of the masses on one hand and cater to rising market aspirations on the other. Unless and until this complex relationship between forestry and outside socio-economic environment is seen through a neutral perspective, any attempt to decide on a course of action for certification of forests in India would be a meaningless exercise. Thus the best way to initiate work on forest certification could be through the establishment of some national initiative. At the start, there is often limited knowledge of certification processes or standards available among the forest industry and various stakeholders.

Many new contentious issues can and will arise, which need to be resolved and worked through in this evolution phase (Handford and Nussbaum 2001). India's forestry sector faces the dilemma of an increasing demand for timber due to burgeoning urbanization and economic growth on the one hand, and the need to conserve, preserve, and restore forest resources for ecosystems and services on the other. As such, there is an ever-growing gap between demand and supply of wood and products. India has a market for secondary processed wood and non-wood products, which has a huge but untapped potential for exports with valorization. The consequent effect could be a benefit sharing among the various stakeholders, including local communities, forest custodians and owners, producers and processors, traders, and retailers. This can be made possible only if an appropriate forest certification system is in place, which also takes care of requisite chain of custody and eco-labelling.

Box 12.2 expounds on the market advantage of certification of forests.

**Box 12.2** Market advantage of certification of forests

In the initial phase, all parties interested in certification should be encouraged to work together from the very beginning to ensure that every one of them, including the general public, have information about certification standards, processes, and their implementation and also to ensure that forest owners, managers, and other stakeholders are able to visualize the pros and cons of certification. There are significant commercial imperatives for forest owners seeking certification, and hence the one who enters first will get an early market advantage, encouraging the entry of others. All this could be possibly achieved through the establishment of a national initiative and subsequent structures and systems.

*Source* Handford and Nussbaum (2001)

## 12.2 NATIONAL INITIATIVE AND WAY FORWARD FOR CERTIFICATION IN INDIA

Establishing a forest certification mechanism in India is now a more urgent need than ever because SFM certification systems need to be encouraged more proactively to achieve the goals of the high priority Green India Mission (as a part of Indian strategy for climate change).

The initiatives taken by the Government of India for facilitating forest certification are discussed in Box 12.3.

**Box 12.3** Initiatives taken by the Government of India for facilitating forest certification

The Government of India has also taken some initiatives for establishing certification systems in the country. In 2006, an expert committee of 19 members was formed for framing the policy guidelines for "Accreditation Criteria and Processes". This committee was given the responsibility to develop an accreditation body for certification in India. The committee recommended some important points for developing an accreditation body in India (Annex 7).

*Source* MoEF (2006)

The Ministry of Environment and Forests (MoEF) has constituted the National Forest Certification Committee (NFCC), a multi-stakeholder committee, for the purpose, which has submitted its report to the ministry in September 2010. The report mainly focuses on the process of designing and delivering the ab initio steps of a credible certification system with a focus on assessing, monitoring, and deploying SFM Principles, Criteria, and Indicators (PCI). It also focuses on the regulatory framework for ensuring the legality of logging and trade, both regarding imports and exports and also for promoting private–public partnership for contributing to SFM, forest restoration, green initiatives, Reducing Emissions from Deforestation and Forest Degradation plus, payment of ecosystem services for developing and designing a national forest certification system and installing an Indian Forest Certification Council. The main recommendations of the committee are included in Annex 9.

The MoEF has constituted a core committee to analyse the recommendations of the committee report and discuss it with various groups having stakes in forest certification to have a better understanding of all related issues and develop mechanisms to address the ones that are most urgent. The outcome of consultation is given in Annex 10. In June 2011, the core committee and various stakeholders (the representatives of industries and some other stakeholders) discussed and agreed in principle to establish a national-level forest certification council (Indian Forest Certification Council-proposed). This engagement was critical as international experience has shown that pushing forward with standards without general support and involvement seldom gives a good result. The overall support from key industry and stakeholder organizations has been low in India initially but is essential while moving forward (Handford and Nussbaum 2001). The draft composition and eligibility criteria are given in Annex 11.

In the process of developing a mechanism/framework for forest certification by the MoEF, Government of India (various crucial issues were identified and addressed in the process of developing certification activities both for natural forests and plantations outside the notified forest area along with non-timber forest products (NTFPs). Some of the issues addressed so far are as follows.

1. Development of appropriate national regulatory, fiscal, and institutional structures that are essential for the certification system to operate and the mechanisms to implement the certification schemes.
2. Participation of stakeholders in setting of national and regional standards and in the implementation of the certification processes including smaller growers/farmers and communities.
3. Role of a nodal/national autonomous agency for forest certification in India.
4. Ensuring credible and feasible certification schemes encompassing the elements of quality of assessment of forest management and acceptability to all key stakeholders.
5. Cost effectiveness and commercial viability.
6. Ensuring comparability and equivalence in standards with other international certification imitative and in consistent with the provisions of World Trade Organization.
7. Role of Government of India and state governments in developing policy and legal framework for certification activities and also in developing the certification process.
8. Establishing accreditation or licensing standards for 3rd party certifiers and recommending mechanisms for registration, monitoring the work of certifiers.

9. Raising of awareness about forest certification amongst cross section of state or national stakeholders.
10. Clearly defining the objectives, area and responsibility to bear large expenses of forest certification.
11. Developing certification capacity in civil society and imparting of training and technical assistance etc.

During multi-stakeholders discussion by the National Forest Certification Committee and also by the Core Committee of MoEF, most of the issues mentioned above have been taken care of and a conscious decision to establish a forest certification council at the national level has been taken. The framework of the Indian Forest Certification Council has been drafted by the Ministry and is in the process of finalization.

Before putting the Indian Forest Certification Council into place, the task before the government or the council (after its establishment) is to identify and analyse the following critical issues and possible options.

- The identification of important beneficiaries of the certification process.
- The current requirement of "certified forest produce" in terms of volume with respect to the reasonable future period (both for domestic consumption and also for meeting export demand). Unfortunately, there are no such market survey studies available for India. In the absence of such market demand surveys for certified forest products, it would be a futile exercise to make any concrete strategy.
- Based on the above assessment, deciding whether to have a "national standard" or whether to leave the "forest owner" to choose any other "forest standard" available in the market.
- The decision made would be based on the "volume" of "certified forest product" required as a "national standard" or Indian Forest Certification Council can only be functional if there is sufficient volume of business to make it economically viable.

We already have the example of Forest Stewardship Council (FSC), which claims that it is still struggling with economic viability issues despite so many factors working to its advantage, such as huge market support, first mover advantage in market, support from many environmental non-governmental organizations, maturity and experience, and better image and credibility. Hence, if certification agencies like FSC are not able to reach self-sustaining mode, it would be difficult to visualize the future of an Indian National Forest Certification Agency, which could be facing a lot of handicaps, at least in the initial years.

- Even if the Government of India or state governments ignore the above-mentioned issues and feel concerned about the expansion of forest certification, either for "sustainability" or "profit to growers", it can be facilitated by initiating the proposed National Forest Certification Council, which may need financial aid in the initial phase.

However, there is a strong argument which may discharge the government from adopting this option. One of the fundamental requirements of any credible certification process has to be that it is independent. Hence, government support and thus control (of any kind or scale) would be undesirable for the credibility of the scheme and may thus defeat the very purpose of having it. This in future could become a major factor in its becoming unpopular. Forest certification is always considered a market-driven approach. This is evident in the flow chart in Figure 12.1.

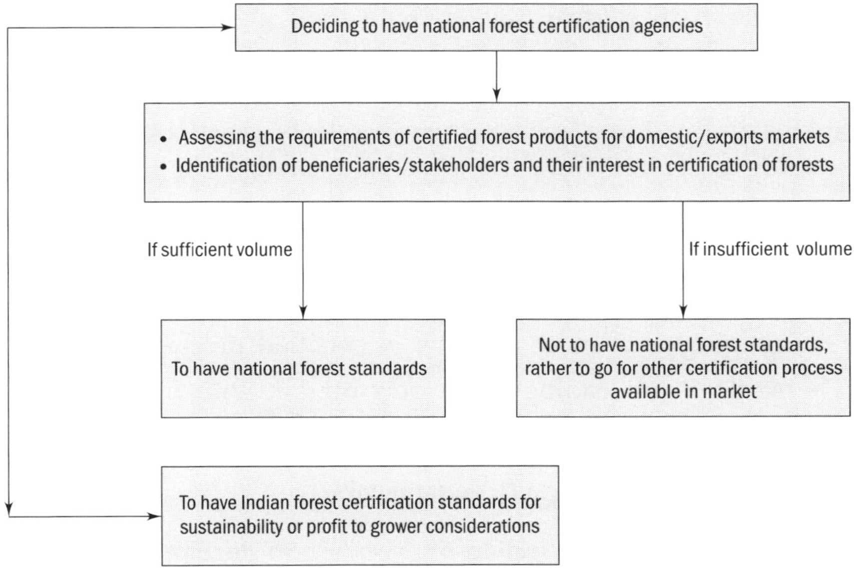


**Figure 12.1** Framework of forest certification in India: possible options

- If the option decided upon is to have something like the National Forest Certification Council, not dependent on and driven by the government, all the beneficiaries of forest certification have to come forward. In the first instance, these could be plantation owners (particularly the large ones, which may include forest development corporations also) and users and exporters of certified timber/non-timber forest products, which appear to be major beneficiaries and have financial capability also.

  It is desirable that they should take active lead (as they are the first ones to get assured financial returns) and help in formation/ evolution of national forest certification standards and processes. After this stage, more and more stakeholders can and should be roped in to make the whole standards setting and operational part more participatory and widely accepted. Of course, in the very initial phase, the government may take the lead in the process through a national certification initiative. The process is depicted in Figure 12.1.
- Once it is assumed that the National Forest Certification Council is going to be set up, it becomes very important to consider what national standards should be. This standard setting has to be done in the wider context of different standards of forest certification available in the global market (with their given client focus and cost of operation, as major considerations).
- As forest certification is envisaged as a market tool for ensuring sustainability, India as a country has to be very clear on what it intends to achieve from this process, including the following.
  - Using this concept for ensuring sustainability of forests.
  - Helping growers and intermediaries/processors in getting more value for their effort.
  - Emerging as a major exporter of certified produce.

The strategy for leading the export market with Indian certified produce is explained in Box 12.4.

## 12.3 SFM AND CERTIFICATION IN INDIA

We will now analyse the Indian effort made in developing standards of SFM initiated by the Indian Institute of Forest Management (IIFM), Bhopal. Many workshops and meetings have been organized by the

**Box 12.4** Strategy for leading the export market with Indian certified produce

If the country wants to reach out to the international market on the strength of its forest certification brand, it would require a different strategy. This strategy could be much more complex as it must be in a position to cater to the requirement of many and diverse markets.

- If it wants to cater to or feed the global export market only selectively, the job is slightly easier because forest certification standards would be attuned to the requirements of a limited number of markets.
- If it does not intend to enter into the export market but targets the domestic market for certified forest products, the job is easier. In this case, the standards and processes can be relatively independent of international requirements. We can afford to have our own country-specific, market-driven forest certification standards.
- If the market study offers evidences that India does not have the capability to export certified forest products in the near future (due to economic scale), we can leave the arena to international forest certification agencies, which could help meet the requirements of a wide variety of forest growers and markets. However, as and when the required capability is felt, we can take appropriate steps for our own standards/certification body.
- The other major reason for having an Indian Forest Certification Scheme and Agency could be to ensure the supply of certified forest products caused by mandatory enforcement of sustainable production of timber. This has already taken place in government procurement (although selectively through the National Building Code and green procurement policies) and gradually can be applied to all private procurement also. In fact, concepts such as green building initiatives are going to be major drivers for such a shift and the consequent demand for certified forest products. This way we can support the global pressure to practice SFM at the national level by raising the demand for such products. However, this would be termed command and control or regulation-driven movement and would lack the voluntary quality.

IIFM at national, state, and regional levels to popularize and sensitize different stakeholders to this idea. The outcome of such efforts is seen in the shape of a broad set of criteria and indicators for SFM in India. This has been adopted as a "national draft set of criteria and indicators for India". Now this set of criteria and indicators by the MoEF needs to be finally adopted, taking in many considerations, including the market-driven factors mentioned above. This is not an easy job for the Government of India, given the complexity of issues involved and the objective purpose of the whole exercise.

However, it can be said that even this approach could help in earning India a clean or environment-friendly image at the global level. But the flip side of this regulation-driven exercise is that it would not be able to leverage the efficiency and cost aspect, which is the case, when the whole exercise is market driven.

## 12.4 SHAPE OF SFM AND FOREST CERTIFICATION: ROADMAP FOR EVOLUTION

The ultimate shape of SFM in India should be guided by the following two considerations.

1. Requirements of market (assuming many scenarios).
2. The final shape of general SFM should come out clean in the eyes of the global community. If ensuring sustainability in forestry is the only criteria, issues like institutional capacity of the forest department, ensuring a smooth transition to changed management practices, and requirements of finance and expertise would be important considerations for a future course of action.

Given the limitations of finance, lack of initial expertise and appreciation of the general reluctance/resistance for abrupt change in a system, taking a precautionary or guarded/limited approach (instead of attempting a quantum jump) would be prudent for India. Thus the current situation clearly indicates that we go for change in a phased manner.

Prudence also dictates that we assess and capitalize on our strength, that is, the existing institutional framework and expertise of foresters who are quite comfortable in managing the forest as per working plans. Hence, the existing working plans can and should be the initial platform for this change. By adding some extra requirements of sustainability to them, it would be very possible to make working plans conform to SFM requirements. In fact in one study (Yadav, Kotwal, and Menaria 2008), it is reported that 70%–80% of criteria and indicators are somehow assessed in the current shape/structure of the Indian Working Plan Code. Hence, to start with, if India initiates its SFM efforts by improving the current working plan itself and by suitably making minor changes in nomenclature, process, and content, it could be the first phase of Indian SFM.

This approach would neither require many resources nor face much resistance and psychological barriers with foresters. At the same time, the country would be in a position to show the world that broad issues of sustainability have been incorporated in general forest management in India. At the same time, the country would not be accused of not doing anything concrete towards SFM as many critics currently think. The world will see that the changed working plan would be highly oriented towards SFM and be able to achieve SFM by incremental changes.

In three–four years, it may be possible to add more criteria and indicators that may require less external expertise and fewer resources and, in a phased manner, attain full-scale SFM at par with global standards. The proposed evolution of SFM can help in ultimately evolving to and matching the standards of SFM and forest certification at par with global standards. Although we have visualized the scenario in isolation, thus making it very simple for ease in understanding, the reality is quite complex, but still manageable.

By adopting this approach for SFM and certification of forests in India, we can increase stringency and rigour over time as the scope and demand for certification increases. We can address the most urgent environmental and commercial issues from the very beginning but in a phased manner. This would be not only institutionally workable but also cost effective and help in improving efficiency.

However, the exact market requirements of certification based on research/survey—as mentioned in an earlier part of this chapter—would be able to provide a clear picture and subsequent options. Only then the appropriate shape of SFM and forest certification standards can be decided at the national level.

## 12.5 CONCLUSIONS

In all probability, the reality will require that some consideration of market and of sustainability will be needed. Hence, the final shape of SFM and its criteria and indicators will be decided by both these requirements.

13

CHAPTER

# Challenges of Certification

## 13.1 IMPORTANT FACTORS LIMITING THE SPREAD OF CERTIFICATION

The certification of forests has got a promising role in the spread of sustainable forest management. But many issues still remain to be addressed if certification has to gain a real momentum in the future. Some of the major issues and emerging trends in global forest certification are discussed in the following sections (Vogt, Larson, Gordon, *et al.* 2000).

### 13.1.1 Slow Progress of Certification in Developing Countries

Forest certification was initially devised to regulate the menace of tropical deforestation. However, the vast majority of certifications of forests at present occurred in Europe and North America, while developing countries (which have most of the tropical forests) contributed a mere 8% to the total certified forests till 2002 (ITTO 2002). Analysing the principle cause of the disparity in distribution in certified areas between developed and developing countries, it is important to note that many developed countries have a longer track record of supporting and enforcing regulations that support sustainable forest management. Thus, the overall direct impact of certification on timber-producing tropical countries has been very little till date, which is a matter of serious concern. Certification as a tool has not worked or rather has not been accepted for forests in the region for which it was initially meant.

Several factors can be attributed to this situation, as identified by Atyi and Simula (2002).

- Inflexibility of certification standards.
- Failure to recognize broader local land-use issues.
- Conflicts and incompatibility between legal settings and certification standards.

Developing countries are positioned quite differently compared to developed countries with regard to their certification needs and possibilities as well as resources they have for making use of certification. Tropical timber producers are far more concerned about the economic aspects of certification such as expected increase in production costs, uncertainties over market benefits, and difficulties in achieving certification status. They consider that forest certification is more a market requirement imposed by importers, which is difficult to comply with, and see this as a trade barrier rather than an aid for promoting their exports (Durst, McKenzie, Brown, *et al.* 2006).

### 13.1.2 Limited Markets for Certified Products

North America and European countries have emerged as major markets for certified wood products. But there is little or no local demand for certified forest products in developing producer countries at present, nor in the major importing countries of Asia. In these countries, most demand for certified products currently comes from businesses and government agencies wanting to pursue an appropriate environmental policy towards sustainability (Durst, McKenzie, Brown, *et al.* 2006).

The certified products initiative in the USA and Europe is led by "buyers groups" of forest product retailers and traders, especially retail home-improvement chains. Lowe's and Home Depot in the USA, UK's B&Q, Home Base, Sainsbury and Meyers (members of "1995+ Group") are some of the most significant members of such buyers groups. However, compared to American and European markets, certification has not gained any meaningful market share in principal Asian markets such as Japan, Korea, and China, which are among the leading tropical timber importers.

Research also indicates that there is very little recognition or demand from private end consumers, who are generally unwilling to pay extra for certified forest products (Anderson and Hansen 2004), except for certain niche products such as high-value furniture and musical instruments.

Thus there is insufficient demand for certified products in global markets. Certified timber has not delivered the price premium that was initially expected.

### 13.1.3 Issue of Private Non-industrial Forests

Most certification of forests at present has happened for industrial forests and plantations. In India, a smaller proportion of privately owned forests are owned by individuals/communities or non-governmental institutions, and expecting certification of such forests would be farfetched, mainly due to economic reasons.

The failure or lack of willingness of most forest certification protocols to look at patterns and processes at the landscape level has undermined the comprehensive role of forests and has been an important factor in poor acceptability of "certification" as a concept.

### 13.1.4 Wide Gap Between Existing Management Standards and Certification Requirements

In many developing countries, there is a wide gap between the existing standards of management and what is required by certification schemes. This problem gets aggravated by the fact that there are often insufficient financial and human resources to effectively raise these standards (Durst, McKenzie, Brown, *et al.* 2006).

These countries also suffer from a poor ability to formulate appropriate forest sector policies and ensure their effective implementation.

### 13.1.5 Conflicts between Legal Settings and Certification Standards

In most developing countries, there are conflicts between national laws and forest certification standards such as the ownership of land, usufruct rights of forest products and services, sharing of responsibilities between the government and local people,. The forest rights and concessions may be locally defined in a way that does not correspond to the requirements of voluntary certification. What may be considered "illegal" based on public law or unacceptable for voluntary certification standards may still be part of customary law and traditional rights of local people living in and around the forests.

There is still no harmony between customary law and traditional rights of local people living near forest fringe areas and national laws/acts dealing with forests and land use."

### 13.1.6 Difficulties for Small Producers

By and large, forestry producers who have embarked on certification are basically large companies located mostly in industrialized countries and already locked into the ongoing international sale of wood-based products, which have not significantly changed their forestry practices because they were already performing to adequate standards (Thornber 1999).[1] For producers who do not follow sustainable forestry practices and which are generally smaller in size and cater towards a more local market, certification has not accorded any real incentives to modify their behaviour because they see no added premium to the price of certified wood (Cerda and Lira 2002) on the one hand and forest certification means less profit than before (Howard, Rice, and Gullison 1996) on the other.

In case of tropical forests, increasing requirements for heightening quality forestry management will not help conserve these ecosystems because a large part of land use is not driven by objectives like maximizing forestry returns but by its alternative value when used for agricultural purposes. This leads to "slash and burn" to clear the land and quickly change the ground's productive capacity (Pearce and Moran 1994). Therefore, the diminishing of the profitability of forestry management through certification stands in the way of real opportunity costs presented by alternative uses, which are, in many cases, less sustainable than those they sought to correct.

### 13.1.7 Doubtful Side of Certification

While the system has managed to produce a small supply of certified timber, doubters continue to be vocal about a number of important issues.[2]

- One of them is the question whether there is any real demand for certified "green premium" wood. Market evidence, to date, suggests that there is not.

[1] Details available at <www.fscoax.org>

[2] Details available at <www.forestry.about.com/od/certification/Forest_Certification.htm>

- Certification comes quite low in the list of consumer priorities after price, quality, and fitness for use.
- Another serious doubt often raised is about the credibility of certification processes. The "Teakwood Controversy" is just an example of how reliability and credibility can taint the certification process (Box 13.1). This case highlights how doubtful the process can be when there is no clear mechanism to screen the reliability of potential certifiers or the certified.

The issue of high direct and indirect costs of certification has been separately dealt with in Chapter 7 "Economics of forest certification".

## 13.2 CHALLENGE OF CERTIFICATION: INDIAN PERSPECTIVE

Forest certification as a market instrument to promote sustainable management and production has still not really taken off with regard to tropical forests (for which it was originally intended). Certification processes are still driven by the limited international (that is, particularly Western) market. Demand for certified forest products in national markets of developing countries has hardly grown, even though large volumes of forest products are traded and thus the greatest benefit could be achieved as regards sustainable management. Another reason for this limited growth of certification can be attributed to the direct and indirect costs involved in certification, which is not compensated for in prices (Cerda and Lira 2002).

Evidence gathered so far shows that only large-scale producers from industrialized countries have been most interested in certification, with very few exceptions. The purpose here goes beyond just the issue of price premiums despite certification cost. Certification has discouraged forestry activities in countries where the most important

**Box 13.1** Possibility of fraud

Julio Cesar Centro documents a case where a Dutch-owned company sold economic ownership in a Costa Rican teak plantation where owners were entitled to 85% of the revenues on sales. The certifier of the property was to participate in the profit. Grossly exaggerated claims as to the profit potential were made and a scandal resulted.#

# Details available at <www.forestry.about.com/od/certification/Forest_Certification.htm>

biodiversity is found. Unfortunately, no proper values are placed on preserving the biodiversity of forest resources and they are unable to capture value through certification. This situation then leads to intensifying the problems it sets out to resolve (Cerda and Lira 2002).

If all the above factors are seen in a comprehensive manner, it becomes clear that while the certification of forests is equally or more desirable in tropical forests, their coverage is still poor compared to temperate forests. Similar are the reasons for certification not gaining ground in small and poor countries compared to American and European countries. This is a major challenge to forest certification as a concept and to its progress.

Although certification is well established in other sectors such as food, health, hygiene, organic products, and agricultural products, certification of forests has not achieved such progress. It requires an in-depth analysis that takes into consideration the socio-economic, land tenure rights, administrative, legal, and regulatory factors. Some of these challenges can also be taken as an opportunity to fast forward the process of forest certification in India.

### 13.2.1 Ensuring the Involvement of Different Stakeholders

To take forest certification in the right direction, the development of forest management standards is very important, although developing these standards at a national level is very complex and contentious. The different stakeholders have different and strong views and resist dilution of forest council principles that address their priorities. Certification will not carry credibility unless supported by organizations trusted by the public; so it is vital to involve all stakeholders in the development and testing of standards. There are various conflicts of interests among them, such as local community versus trader versus consumer interest, big versus small operators, north versus south, and global versus national and regional certification systems. The biggest challenge is of achieving unity in diversity.

### 13.2.2 Issue of Site-specific/Region-specific/Country-specific Standards and Processes

Certification is often perceived as a non-tariff barrier in the south. Forests, especially in tropical regions, are very rich in biodiversity, including endemic and rare flora and fauna. If sustainable forest management is considered an important objective for these forests, its imperatives have to be understood and accepted.

### 13.2.3 Compliance to Regulatory Frameworks (National/International)

Mandatory rules, regulations, and conventions set the legal framework of forest certification. Inspections of forest by independent third-party accredited organizations assess the compliance with a set of principle criteria and indicators, including adherence to the legal framework. This practice ensures that timber from certified forests is not only legal but also sustainably produced according to the certifier's criteria. Although certification cannot substitute an appropriate forest policy, a strong legislative framework, and capable forest managers, it cannot have the desired impact in the absence of these. Harmonization of the set of rules/regulations with the principle criteria and indicators is very much required.

### 13.2.4 Issue of Labelling and Logo

Certification schemes need to maintain consistency, credibility, and transparency of standards and auditing for customers and traders. The labelling requirement should be feasible, realistic, and cost effective. At the same time, certification will lose its credibility if it is too undemanding, that is, promoting business as usual or certifying the lowest common denominator.

### 13.2.5 Issue of Costs and Benefits

The costs and benefits of using certified wood products pose a problem. The issue of profitability can be viewed from several perspectives—market share of certified wood products, market growth rate, increased cost of certified wood, small price premium for certified wood products, and lower profit margin for certified wood products, relative to non-certified wood products. The profitability of certified wood products will influence short-term and long-term marketing strategies of companies supplying certified wood products having links with branding. All major certification programmes tend to encash the labels given to wood products, hoping for higher financial returns to compensate certification costs.

### 13.2.6 Consistency and Decentralization

Yet another important challenge facing forest certification programmes is to construct systems that can claim to be globally consistent and at the same time responsive to the diverse local needs of the country. Forest management unit-level certification without national standards can be prone to inconsistent assessment and less support for social benefits, due to differences in interpretation of generic standards by certifiers, which may arguably lower accountability of certifiers hired by forest management units.

### 13.2.7 Expanding beyond Timber

The development of the market and payment for multiple products and services (such as non-timber forest products and products from less known species along with environmental services) are vital to the future of sustainable forest management and timber certification in a tribal country like India. But payment for environmental services could demand higher forest management standards also. In the future, certification could become a major tool for justifying and attracting payments for environmental services. For example, in the context of landscape-scale planning and management, forest certification may have to put much more focus on the industrial forest sector and then on the community forestry sector.

If national certification standards are overly meticulous, inflexible, incompatible with local and customary standards, and ultimately costly for local people (even with group certification), it may be very difficult to access the demanding international market for the above-mentioned products and services.

### 13.2.8 People-friendly Approach and Process

Another important factor is the social aspect, which calls for a concern for people. People, especially those living in or around forests, must have fair share of benefits accruing from the certification process and also may expect to be integrated into the mainstream. Therefore, there is an urgent need for meeting the expectations of these people. Certification of non-timber forest products may enhance the well-being of both forests as well as people, as currently there is a worldwide interest and demand for natural, herbal, and organically grown products. This can and should be seen as yet another opportunity to alleviate the poverty of forest fringe communities of the country by ensuring a better market for the produce collected and processed by them. It is the need of the hour that a people-friendly framework is designed and put in place, which would lead to a situation of "man with forests" rather than "man versus forests". Such a proactive and people-friendly approach by the government will encourage widespread sustainable management of forests.

### 13.2.9 Willingness to Pay Premium

A major drawback in the forest certification concept has been the demand for certification coming mainly from developed countries; it offers benefits only for forest products destined for export to these countries. The consumers of developed countries are not very willing to pay a premium on certified forest products. In India, more than two-thirds of the marketed wood and wood products are consumed locally, where the willingness to pay a premium price for certified products is almost nonexistent. The other main reasons attributable to this situation include lack of information, socio-economic condition, insufficient government support, rigidity of standards, complexities of laws, human rights, and certification standards.

### 13.2.10 Other Cross-cutting Issues

Another challenge to certification is whether it can address the cross-cutting issues of poverty among forest-dependent communities. Ensuring that the forest-dependent communities remain motivated for concomitant environmental services is a prerequisite for forest certification. Forest certification should dovetail in accordingly and, inter alia, get involved in payment for environmental services.

## 13.3 CONCLUSIONS

Governance in the forestry sector is a burning issue with more discussions taking place to address issues of corruption, illegality, and bad governance. It is desirable to change the image of the sector. The well-governed and effective institutions in the forestry sector can prove to be decisive factors for the success of sustainable forest management and forest certification (Cerda and Lira 2002).

Without good governance and effective institutions, the scope of sustainable forest management will remain limited. Good governance includes parameters such as trust and faith in subordinates and the provision of an enabling environment with transparency and accountability. The way participation, roles, rights, responsibilities, and powers are organized between various parties and their institutions, both within and outside the government and within and outside the forestry sector, nationally and internationally, plays a significant role (Cerda and Lira 2002). Greater proliferation of certification in a country indicates the priority given to good governance in the forestry sector. Incidentally, for a country like India, this could be a major reason for the wide practice of certification. This would help India achieve an environment-friendly image in international eyes, in keeping with its emerging status as an important global power centre.

# 14

CHAPTER

# Future of Forest Certification

## 14.1 BOTTOM LINE WITH REGARD TO CERTIFICATION

Over the past few years, forest certification has grown in terms of the total certified land area and also the increasing number of companies that are participating in chain of custody (CoC). These positive trends have continued despite economic challenges within the homebuilding and forest sectors.

Significant changes have occurred within the major certification programmes in recent years, and it is becoming increasingly difficult to differentiate between prevalent certification systems. The challenge is of differentiation in products and services critically important in the marketplace, as consumers find many difficulties in choosing between certified and non-certified products without having to choose between the complexities of individual programmes. For example, some of the dilemmas always haunting a consumer could be in the form of various questions (Fernholz, Howe, Bratkovich, *et al.* 2010).

- Whether a product made of wood from a Forest Stewardship Council (FSC) plantation is more environmentally friendly than the wood from a Sustainable Forestry Initiative (SFI)-certified natural forest.
- Would it matter if a product were from an FSC-certified source as compared to an SFI-labelled product harvested from a forest jointly certified to both SFI and FSC standards?
- Is an SFI-labelled product with 100% certified content more environmentally friendly than an FSC-labelled product whose certification is based on the volume credit system and may not include content from a certified forest of any kind? (Fernholz, Howe, Bratkovich, *et al.* 2010).

There are many more challenges facing the future of forest certification, including higher costs, increased competition from higher performing products, highly competitive markets, and the still unaddressed need to expand the level of marketplace participation to a critical mass. The establishment of a shared approach to CoC systems could be the key strategic activity that has the potential to address multiple challenges simultaneously for major certification systems and for the marketplace. It is a key approach that can generate a win-win situation.

## 14.2 OPPORTUNITIES AND LIMITATIONS OF CERTIFICATION

Despite the slow progress of forest certification, there are several promising opportunities for moving ahead.

- Simplification of certification procedures.
- Phased approach to certification.
- Creating awareness about forest certification and certified forest products.
- Group certification (Box 14.1).

**Box 14.1** Group certification

Group certification is one of the mechanisms to reduce the unit cost and make the certification of small forest ownership and management unit economically viable. It enables small forest owners to join together to reap the benefit of reduced unit cost while maintaining management control of their individual forest. There are many provisions in different schemes and various certification agencies to address this. Some of the schemes have streamlined/simplified the procedures. The Lok Vaniki Act of Madhya Pradesh is an attempt to address the problem of small forest growers/plantation owners, which can be replicated at the national level by means of formulating guidelines/regulations that cover the issues of sustainable forest management and certification as well.

Thus forest certification can offer opportunities for communities to have more control over the marketing and use of the output from the forest. It can provide access to markets outside local communities, with the potential to fetch higher prices and revenues. This may act as an incentive to manage natural resources sustainably.

Besides the merits of forest certification, the concept has some negatives too. Procedural problems have been limiting the proliferation of the concept (Vogt, Larson, Gordon, *et al.* 2000) as has been discussed earlier.

### 14.2.1 Limitation of Certification

- The demand for certified products is small and limited.
- Maintaining CoC from the forest to the customer is challenging.
- Various groups have a stake in different certification systems, which creates both a political and competitive atmosphere among the systems and results in conflicts and divergent claims among supporters of various systems.

## 14.3 FUTURE OF CERTIFICATION OF FORESTS

- The World Forestry Congress held in 2009 in Buenos Aires, Argentina, came out with broad discussions on the future of forestry in the globe, and many of those arguments are valid for the concept of "forest certification" as well. This discussion can provide a very important background for introspection and a guide to a road map for the future. Hence, they are reproduced extensively, with adaptation to the context of certification of forest. All the ideas and points are acknowledged to the World Forestry Congress.
- It is becoming increasingly critical to emphasize that the forestry sector must focus on the outside world and questions and perceptions that are arising there. Traditionally, the sector frequently had a tendency to solve problems by itself for what it sees as its "own" sector. It often views society as the cause of the problems—offering too little support and recognition—and not as an accomplice and facilitator in solving these problems.[1]
- But if we see the complexity of issues in a more unbiased manner, the solutions to the problems of forests have to come from outside the forestry sector through changes in other sectors, society in general, and from political circles. Conversely, major functions that forests perform, if properly managed, for other parties and the cost of losing them have to be highlighted sufficiently. Hence,

[1] Adapted from World Forestry Congress 2009 prescription for forestry

the forestry sector must adopt a more active, strategic, and political position in the public debate regarding forests and should be key to current political and sector agendas, indicating what it actually has to offer. Persuasiveness will be desirable for success.[2]

- New agendas, such as that for sustainable forest management and certification, can help in this regard. At the moment, the necessary skills to operate and communicate strategically are not well developed within the forestry sector. There is a need to make an investment in the skills involved in promoting and facilitating communication, conflict management, consensus-forming, and cross-border collaboration between groups and communities. This entails giving up some of our own (forest or forest sectors') autonomy—or supposed autonomy—and learning to accept being guided by the larger perspective, of which forestry forms only one part.[3]
- There is also a need to give greater recognition to the value of various kinds of processes behind the scenes—processes carried out by a whole range of government bodies, scientific institutes, businesses, and civil-society organizations all over the world, often beginning on a small scale.[4]
- The role of forestry institutions has to and will shift from being one of "do-it-yourself" to one of being a service provider in terms of supplying substantive and policy-oriented forestry expertise, concepts and methods, and high-quality expertise. The aim should be to make forests deliver the best possible contribution to sustainable development while guaranteeing multi-functionality and integrity of forest ecosystems (Box 14.2); sustainable forest management (Box 14.3) and certification of forest are important tools to achieve this. In this context, it is the task of the forestry sector to make clear the actual value of forests in all aspects of production of goods and services, combating poverty, and protecting biodiversity. The sustainable forest management approach could be appropriate in this direction.[5]

---

[2] Adapted from World Forestry Congress 2009 prescription for forestry

[3] Adapted from World Forestry Congress 2009 prescription for forestry

[4] Adapted from World Forestry Congress 2009 prescription for forestry

[5] Adapted from World Forestry Congress 2009 prescription for forestry

**Box 14.2** Holistic view and sustainable forest management

In her address to the World Forest Congress 2009, Seymour wondered: "Can we orchestrate good vibrations?" while referring to the question of what policy and institutions are necessary so that sustainable forest management has a positive impact on local households and society in general. She claimed that this was indeed possible if the forestry sector managed to come out of its shell and made progress in connecting and cooperating with other parties, as a fully recognized and equal stakeholder.

*Source* FAO (2009a)

**Box 14.3** Need for integrity and ethics in the certification of forests

To further add to the uncertainty of forest certification's future, a development movement in the USA recently added a new twist when several legal actions were taken, which are directly or indirectly associated with forest certification. A legal action filed in 2008 was associated with the distribution of softwood settlement monies and the award of some monies to one certification programme without equal consideration of programmes offering what were argued to be of similar benefits or services.[#]

This action included FSC-US as a plaintiff. In 2009, Forest Ethics filed a complaint alleging abuse of non-profit regulations by the SFI and misleading eco-label claims.[†] These claims were filed with the Internal Revenue Service and Federal Trade Commission.

Within weeks of the Forest Ethics filing, forest sector companies responded with a complaint filed against the FSC and the United States Green Building Council regarding eco-label claims of their own for being exclusionary with respect to equivalent certification programmes. Whether any (or all) of these legal actions have merit may be largely beside the point as such activity may serve to simply increase cynicism among public and decision-makers about forest certification and dampen enthusiasm for certified wood use.

[#] Details available at <www.naturalresourcereport.com/2008/10/fsc-sues-us-government-for-helping-aff-and-family-woodland-owners>

[†] Details available at <www.wflc.org/cases/credibleforestcert/SFIcomplaints>

*Source* Fernholz, Howe, Bratkovich, *et al.* (2010)

- The developments and tendencies like ongoing globalization and decentralization, social integration, dynamism, interconnection and complexity, changes in governance, and increasing competing claims on forests constitute major challenges for the forestry sector and are formidable challenges to foresters in terms of functions and positioning.[6]

Hence, in addition to maintaining and guaranteeing substantive expertise, a number of changes are necessary in ideas, attitudes, and methods in the forestry sector if this sector has to remain an interesting, relevant, and effective partner in the development and implementation of the global and local forestry agenda. Sustainable forest management and certification of forests attempt to address these issues, though partially.[7]

Over the past decade, forest certification has enjoyed a fairly steady growth, especially in terms of CoC certification. However, a closer look at the data shows that this trend may be slowing, and there is even evidence that the amount of third-party certified land may have reached a plateau or nearly peaked in 2008. The thought of reaching a plateau in terms of certified forest areas poses serious questions to the feasibility of certifying new lands beyond those that are currently certified. The financial and economic constraints will impact the decisions of renewal of existing certified forest. The economic challenges that many certificate holders face are real, and the threat that existing certificates will be discontinued at the time of annual audits or after five years is very much legitimate. The reality of challenging economic times and the fact that FSC and SFI have changed their standards in the intermediate five years add to the uncertainty.

## 14.4 ROAD AHEAD FOR CERTIFICATION OF FORESTS

At this stage in the development of forest certification, it is important to assess whether forests would be better off with just one dominant system or whether continued competition between programmes offers benefits. In the past, the benefits of competition have been fairly evident. Many of the changes within the FSC and SFI systems over the past six years are the direct fall out of criticisms and comparisons levelled against each and resulting from their competition (Fernholz, Howe, Bratkovich, *et al.* 2010).

[6] Adapted from World Forestry Congress 2009 prescription for forestry

[7] Adapted from World Forestry Congress 2009 prescription for forestry

## 14.5 NEED FOR HARMONIZATION

All over the globe different certification systems are struggling hard to establish themselves in the eyes of consumers. This includes adopting the features of their competitors. One of the important certification systems, SFI, has felt compelled to become internationally endorsed and to establish a CoC programme to stay competitive with FSC. To remain ahead of its competitors, FSC has established independent accreditation and a volume credit control system.

Of late, there is increasing debate about the potential for mutual recognition between certification programmes. The Programme for the Endorsement of Forest Certification (PEFC) approach to endorsement offers an example of how mutual recognition can work. FSC's use of national initiatives and standardized principles and criteria are yet another model for replicating certification standards on a global scale. While competition among forest-certification programmes is likely to provide ongoing incremental benefits, it is not clear about the benefits of competing CoC systems. Whereas competing forest management standards illustrate some fairly prominent differences in defining responsible forestry, the CoC standards are much less distinct. Hence, it would be desirable for the various certification programmes to reach an agreement on a common approach to CoC that would allow standardization throughout the marketplace. In this way each programme could still retain distinct differences in logo and trademark usage regulations, although the actual details of control systems, record keeping, training requirements, and the exclusion of unspecified risk or controversial sources are already highly compatible. If all the programmes could agree to a common, standardized approach to CoC requirements, it might allow for more widespread and rapid adoption in the supply chain. CoC standardization might also allow each programme to refocus energy and resources on other more important aspects of their work, which can be the refinement of forest management standards and expansion to engage a larger percentage of the world's forests (Fernholz, Howe, Bratkovich, *et al.* 2010). This issue assumes great significance in the Indian context as efforts to initiate the national certification standards and processes are under way.

## 14.6 FUTURE THREATS TO CERTIFICATION OF FORESTS

One of the significant threats to the future usefulness of forest certification may be the development of product-specific standards. So far forest-certification programmes have taken a holistic approach to natural resource management, inclusive of all types of forest-derived products and benefits. Simplified approaches that focus on single issues and single products may prove to be highly competitive and undermine the market demand and customer commitment to general forest management certification. Ideally, to remain competitive, forest management standards should be in a position to incorporate specific indicators to address emerging issues and opportunities, thereby retaining an edge and maintaining a commitment to comprehensive and systems-based certification (Fernholz, Howe, Bratkovich, *et al.* 2010). Again the issue raised here is important from the Indian perspective as it may help in providing direction to the proposed national certification process.

## 14.7 ENHANCING THE EFFORTS OF SUSTAINABLE FOREST MANAGEMENT AND FOREST CERTIFICATION

A balanced analysis calls for a recognition that forest certification is not the only, nor necessarily, best solution to the problems faced by forest sustainability. However, on the basis of experience, five elemental policies that could enhance the performance of this tool are important, assuming that its use will supposedly continue to grow the forest products market (Cerda and Lira 2002).

### 14.7.1 Pragmatism

In the search for sustainable forest management, the temptation to seek perfection has to be avoided at all costs (Spears 2000). When the requirements of the different standards are examined, not in light of their principles but in light of their respective criteria and indicators, the existence of diminishing marginal returns tends to be forgotten within the framework of their effectiveness. As has been mentioned extensively throughout, such a situation runs the risk of leaving marginal groups operating in the more "sensitive" areas out of the picture. A halfway solution would be to differentiate the requirements for certification over time in accordance with the size and importance of producers, establishing certain requirements as basic, such as harvest not exceeding the growth of the resource in the proper space of time or the obligation to ensure the reforestation of areas harvested (Cerda and Lira 2002).

### 14.7.2 Fairness

With the present dynamics of forest certification, it can never be financed by marginal producers unless they are helped by external financiers or regulatory changes are introduced into the existing certification schemes (Markopoulos 1998, 1999a,b). The north–south syndrome of impartiality, so intensely lauded,[8] requires a period of adaptation. The solution to different requirements according to the type of producer as mentioned earlier, in conformity with the logic that continuous improvement is important in many certification schemes, could indeed provide proper incentives for resolving the more urgent and serious problems of forest sustainability (Cerda and Lira 2002).

### 14.7.3 Focus

Scaling certification requirements whilst helping marginal producers to take part in sustainable forest management practices would also help to detect by default those who operate below the minimum standards required and, as a result, to focus better on other policy tools for regulating the use and conservation of forests (Cerda and Lira 2002).

### 14.7.4 Representative Participation

One of the most distinctive characteristics of the existing certification schemes is the need to validate its credibility by means of proper citizen participation. This participation is channelled basically through the incorporation of environmental non-governmental organizations (NGOs) into the different areas where the standards are created and supervised. Such a practice is reasonable in industrialized countries, bearing in mind the number and variety of existing organizations. However, in underdeveloped countries, environmental NGOs are not only few in numbers but also new to the business, which does not allow them to cover the wide spectrum of preferences and

[8] Details available at <www.fscoax.org>

interests of the population at large. So the certification process so far is noninclusive with the consequent risk of replicating external agendas, which only coincide by accident with regard to local realities. For this reason, representative participation becomes more important within any certification process in developing countries, including India (Cerda and Lira 2002).

### 14.7.5 Concomitance

The other side of the situation described earlier is that faced by developing countries in the Third World where government supervision is weak or non-existent and where the role of environmental NGOs assumes significance, and in many cases, an irreplaceable substitute for achieving forest sustainability.[9]

## 14.8 NEW DRIVERS PROMOTING CERTIFICATION OF FORESTS

### 14.8.1 Overview

In addition to the challenge of declining public-sector budgets and continuing costs of auditing and compliance, expecting growth in market demand for certified forest products represents another challenge to the certificate programmes. There are four main drivers of market growth for certified wood products. Although interrelated, these have differing dynamics and impacts on different sections of the forest products industry—mainly, paper (publishing, printing, and packaging), green public procurement, green building, and legislation against illegal logging (Oliver 2010).

#### *14.8.1.1 Paper, publishing, printing, and packaging*

The attitude of large publishers and other customers of the paper sector has probably been the most significant factor driving growth in forest and CoC certification. For example, Time Inc.'s commitment to increase the use of responsible paper sources (including certified paper as well as increased recycled content paper), beginning in 2004 with targets for suppliers to comply with by the end of 2005, played an important role in encouraging certification activity throughout the major paper-supplying regions of North America (Oliver 2010).

[9] Details available at <www.economist.com>

This growth of CoC certificates is also evident. For example, at least one-third of the 3645 FSC-CoC certificates issued in the USA by May 2010 were given to printing and publishing companies, while a further 20% were engaged in the supply of pulp and paper products. Similarly, 80% of the 976 SFI-CoC certificates issued by May 2010, and almost half of the CoC certificates issued by PEFC, UK, were in the paper and packaging sector. However, this impact has not percolated to the forest regions lacking production facilities for paper (Oliver 2010).

In India also, many paper majors like ITC and BILT have gone for CoC certification in a major way and they use it for the export and premium segment of the market. This trend is spreading to other paper industries operating in the country.

#### 14.8.1.2 Green public procurement

Green public procurement policies for forest products are a relatively new instrument and their implementation is still at an early phase in most countries (Box 14.4). By May 2010, at least 12 national governments had operational green procurement policies, including specific criteria for forest products. These include eight countries in Europe—Belgium, Denmark, France, Germany, the Netherlands, Norway, Switzerland, and the UK.

**Box 14.4** Public procurement policies—turning green

Although the number of national governments adopting green timber-procurement policies has not increased since January 2009, there have been significant developments at local levels. For example, the Olympic Delivery Authority has specified that PEFC- and FSC-certified timber will be used for all construction, including buildings, transport, and energy infrastructure, for the 2012 London Olympics. In Russia, the Ministry of Natural Resources recently endorsed "green standards" for the 2014 Winter Olympics in Sochi, which specifies that FSC-certified timber will be used.

*Source* Oliver (2010)

With the exception of Denmark, most of these policies are mandatory for central government authorities. They are also being promoted to local government agencies. Most of the existing policies go further than European Commission guidance (which proposes that timber, at a minimum, should be from demonstrably legal sources) by establishing a minimum requirement that all wood must be verified as sustainable (including in the UK, the Netherlands, Belgium, and Germany) (Oliver 2010).

Although several other countries are in the planning stage, the list of countries that have operational public-sector timber-procurement policies has changed little since January 2009. However, developments over the recent past have mainly been concerned with moves to strengthen and broaden the policies already in place (Oliver 2010).

For example, the UK government recently expanded the scope of procurement requirements to include social issues alongside the existing environmental requirements. The minimum requirement of the Netherlands government that all wood must be verified legally has later been upgraded to a requirement for verified sustainable wood from 1 January 2010 onwards. In France, the government commitment to promoting increased wood consumption is matched by a requirement that 100% of wood purchased in government contracts must be "legal and sustainable" by 2010. The government recently enhanced its commitment by amending its existing relatively flexible definition of "legal and sustainable" timber and defining "the modalities for recognition of forest management certification schemes" (Oliver 2010).

There are significant differences in the detailed legality and sustainability requirements of government procurement policies. This is a concern for timber producers supplying several markets. On occasion, consumer country governments, responding to the wishes of their domestic stakeholders, have demanded that amendments be made to international certification standards and procedures before acknowledging sustainability credentials. This creates challenges for international certification frameworks when it is necessary to comply with internal rules and timetables for review of standards and when a global consensus must be built for almost every change (Oliver 2010).

Despite these problems, the governing bodies of both FSC and PEFC frameworks have, in most cases, been successful in ensuring that their frameworks are accepted as conforming to the highest sustainability standards. As a result, achieving certification to both FSC and PEFC standards is the most effective way to overcome the potential barriers to trade presented by the diversity of national timber-procurement policies (Oliver 2010).

However, this concept is not completely in tune with the principle on which certification works, that is, voluntary compliance to sustainable forest management and letting the market provide incentives. "Green public procurement" believes in the "command and control" approach and tries to create a market for certified products by force of the state. The move to develop comprehensive public-sector timber-procurement policies has progressed more in Europe than in other subregions of the globe.

Box 14.5 discusses the efficacy of green public procurement policies.

**Box 14.5** Efficacy of green public procurement policies

Recent market research undertaken by the Forest Industries Intelligence Ltd for the UK Timber Trade Federation and Department for International Development suggests that, to date, public-sector procurement has had only a limited impact on timber-procurement practices in European Union member states. The policies directly affect only a small proportion of the overall timber trade, and their effectiveness is undermined by inconsistent application between and within European Union member states. So far, only the governments of the UK and the Netherlands have followed up implementation of the policy with systems of monitoring. However, there are also indications that, with sufficient political will and resources, the influence of government procurement policies can, in time, extend well beyond the direct impact on immediate suppliers and beyond. For example, such policies can increase the sensitivity of larger importers, merchants, and manufacturers to negative publicity.

*Source* Oliver (2010)

The governments of Japan and New Zealand have also developed comprehensive timber-procurement policies. Similar guidelines are also being developed and implemented in several other countries with the potential to make an impact on the demand for certified forest products, including in Australia, the USA, and China (Oliver 2010).

#### 14.8.1.3 Green building initiatives

Unlike the "green public procurement" concept, which believes in using the direct power of the state to create market demands for certified products, "green buildings initiatives" may use the powers of the state indirectly. This can be more successful when the initiatives are exclusively framed and promoted by the state. However, this is not always the case. Such initiatives may be promoted by voluntary efforts (without any support/contribution of the state); it can then be called a non-state-driven, market-based initiative.

The construction sector, which is estimated to account for over 50% of global carbon emissions, has become a key focus of government policies to tackle climate change. Economic stimulus packages in recent years, both in the European Union and the USA, have earmarked funds for green building. The number of operational or planned national green building councils, or organizations established to promote green building practices at the national level, increased from less than 10 in 2007 to 66 in early 2010 (Oliver 2010).

There are examples of Green Globes (USA and Canada), BRE Environmental Assessment Method (BREEAM) of the Building Research Establishment (UK), Built Green Canada, Built Green Colorado, Comprehensive Assessment System for Built Environment Efficiency (CASBEE) (Japan), and the National Green Building Standard of the American National Standards Institute (ANSI) (USA), which recognize multiple forest-certification standards, including FSC, PEFC, and SFI, whereas the Green Building Council of Australia recently ended an FSC-only preference.

In practice, green building programmes have fallen a long way short of fulfilling their potential to drive increased market demand for certified wood products. There are a number of reasons. Although growing rapidly in some countries, the increase in GBIs is still in the early stages. Furthermore, in nearly all systems, the credits available for materials selection make up only a small proportion of the total credits available and are typically overwhelmed by credits for measures to improve the energy performance of buildings during operations. The challenges associated with obtaining material credits often result in design teams grabbing for alternative credits instead. A broader problem is that some green building systems, including Leadership in Energy and Environmental Design (LEED) (Box 14.6), do not adequately consider life-cycle assessment in material specification, and this puts all timber—irrespective of whether certified or not—at a disadvantage compared with other materials (Oliver 2010).

**Box 14.6** Green building initiative, LEED, and the debate

While most green building initiatives (GBIs) now credit wood certified through a range of frameworks, controversy continues to surround the approach adopted by the Leadership in Energy and Environmental Design (LEED) framework, which remains the dominant system in the USA. To date, LEED has offered a single credit for FSC-certified wood products and no credit for wood certified to any other forest-certification system. A review of LEED's timber certification credit initiated in 2008 led to drafting of new proposals for the credit so that it no longer includes a direct reference to FSC and instead ranks certification systems against the US Green Building Council's "Forest Certification Benchmark". However, the review process seems to have left all interests dissatisfied. FSC has been pushing hard to maintain the concept of an FSC-only credit. Advocates of alternative certification systems maintain that the draft benchmark—a third version of which was issued for public comment in February 2010—is so closely aligned to the existing FSC framework that certification systems from different cultural and technical perspectives will continue to be discriminated against.

While the outcome of this debate may have a significant long-term impact on the future market share of different certification brands in the USA, it should not be allowed to obscure wider issues surrounding the use of wood in the various green building programmes.

*Source* Oliver (2010)

The current focus on energy efficiency in construction suggests that there is huge potential for growth in GBIs with significant implications for the growth in market demand for certified forest products. The USA-based LEED and the UK-based BREEAM are probably the most well-established GBIs. However, such initiatives are now proliferating with efforts under way in many countries to develop new nationally adapted programmes; systems like CASBEE in Japan, High Quality Environmental Standard (or Haute Qualité Environnementale, HQE) in France, German Sustainable Building Council (or Deutsche Gesellschaft für Nachhaltiges Bauen, DGNB) in Germany, and Green Globes in North America are gaining momentum. These standards are now making strong progress in some regions not previously regarded as particularly "green". For example, the capital of the United Arab Emirates, Abu Dhabi, has set its sights on enacting the world's toughest green building standards. While the potential is there, evidence suggests that a considerable amount of spadework is still required both to increase uptake of GBIs and to ensure that standards give appropriate credit to wood's environmental attributes. In fact, GBIs can be a mixed blessing for certified wood products (Oliver 2010).

Wood can actually be discriminated against by GBI standards, as wood is often the only material required to demonstrate responsible sourcing. GBI standards giving exclusive recognition to particular forest-certification brands may help drive demand for these brands at the expense of wider appreciation of the environmental merits of wood.

But there are signs that some key consumers of GBI standards are becoming disenchanted with the concept for several reasons (Oiliver 2010). At the United Nations Economic Commission for Europe (UNECE) Timber Committee Workshop on Green Building Initiatives (2008), representatives of the Architects' Council of Europe suggested that the concept behind many existing GBIs was deeply flawed. They noted that it is quite possible to manipulate credit systems to design a building that is not very environmentally sound although achieving a high rating. Poorly designed GBIs can reward building planners for taking a few environmentally progressive steps, some of which may not be particularly relevant, while ignoring deeper problems. Concerted efforts are being made in various forums with the aim of overcoming these problems and of improving the application and conformity of GBIs. For example, the International Organization for Standardization (ISO) is now considering GBIs under Technical Committee 59 on sustainability in building construction. In the European Union, the CEN Technical Committee 350 is working on sustainability of construction works (Oliver 2010).

#### 14.8.1.4 Other issues

The certification of forests must be socially useful also. They should have a positive bearing on the livelihood aspects of communities living in and around the forest.

The initiatives taken under the Reducing Emissions from Deforestation and Forest Degradation programme can prove to be an important driver for forest certification as it is essentially an attempt to promote sustainable forest management. Like forest certification and payment for environmental services, it works through financial markets to provide economic incentives to forest managers and stakeholders in natural, tropical forests.

## 14.9 CONCLUSIONS

It may be too early to assess the impact of forest certification as the concept is still evolving. But some of the positive outcomes of the last 20 years show that forest certification has led to an increase in demand for timber products from well-managed sources (Box 14.7). It has slightly improved forest management practices—mainly in the north, and working conditions—mainly in the south. It has not solved the basic problems, but the problems have been highlighted effectively. Another positive outcome is that it has increased the understanding of what constitutes good forest management that needs to be developed jointly by economic, social, and environmental interest groups.

**Box 14.7** Legal timber

The legality of timber is reflected in the three pillars of sustainable forest management besides being ensured by a process involving:

- The granting of legal rights to harvest.
- Compliance with forest management regulations, including relevant environmental and labour laws.
- Taxes, fees, and levies.
- Respect for other parties' tenure/use rights.
- Trade and export procedure compliance.

*Source* Abdullah, A. L. 2011. Personal discussion on 31/3/2011 at FLEGT, Malaysian office, Kuala Lumpur

The challenges have been enormous given the complexity of forest management, especially in the face of serious problems like weak governance, poorly defined tenure rights, and institutionalized corruption. The major economic problem has been that while there is demand for sustainably harvested forest produce, the market has largely not been willing to pay higher cost for sustainable management. Even then there is not much pressure from the stakeholders directly affected by certification, and the government is not taking any proactive measures although having more than 95% stake as owner of the forest. The current political will is limited and is not triggering development of credible standards and criteria through national initiatives that put pressure on producers to seek easier standards of sustainability, which allow them to tap premium markets. Certification also faces a serious dilemma between expanding the supply of certified products to meet the emerging demand and maintaining a credible standard that takes into account all aspects of forest management.

Of late, the focus has been on establishing standards for forest managers and developing a critical mass of certifiable timber. However, the focus has to be on strengthening national institutions, policies, and legislation to reduce the gap between current levels of forest management and certification requirements, reducing the cost of certification, better access to markets and incentives for trade in certified forest products, combating illicit felling and illegal trade, promoting payment for environmental services, linking the issues of reducing emissions from deforestation and degradation in forest certification, and thus increasing the effectiveness of the concept.

# Annexes

## Annex I

### INDICATIVE STRATEGY FOR ACHIEVING FORESTRY AT THE NATIONAL LEVEL

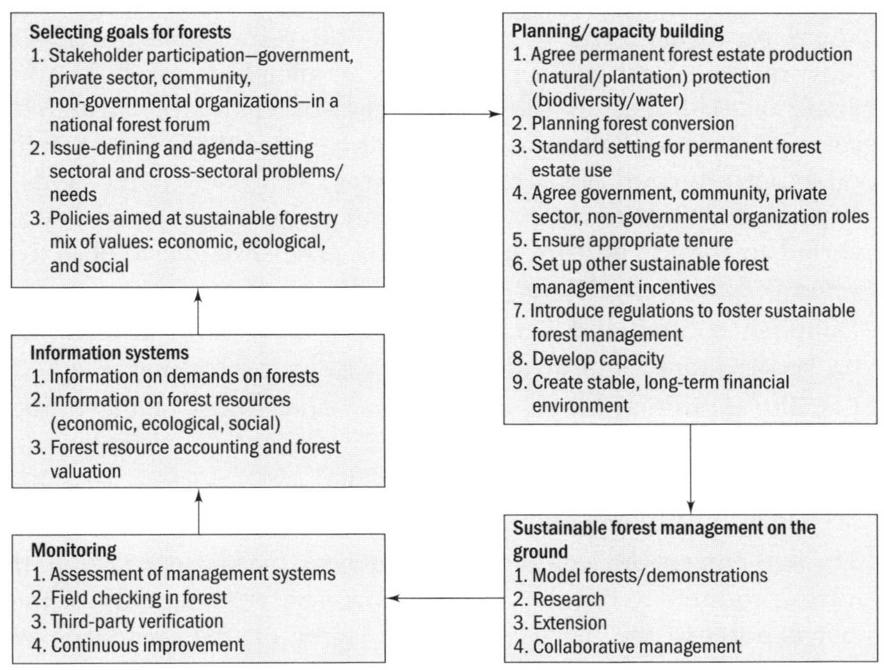


**Source** Upton and Bass (1995)

# Annex 2

## EUROPEAN UNION TIMBER REGULATION/VOLUNTARY PARTNERSHIP AGREEMENTS

The European Union-Forest Law Enforcement, Governance and Trade (EU-FLEGT) process started in 2003 as a response to political commitments made by the EU, member states, and partner countries as part of the G8 Action Programme on Forests and the ministerial conferences on Forest Law, Enforcement and Governance (FLEG) facilitated by the World Bank.[1]

The EU-FLEGT aims to develop markets in Europe for legal products by focusing on trade policies and responsible purchasing by governments and timber importers in member states. The key measure of EU-FLEGT is the development of Voluntary Partnership Agreements (VPAs)—bilateral agreements between the EU and tropical wood-exporting countries. The aim of these voluntary agreements is to improve forest governance and provide guarantees that the wood imported to the EU is from legal sources. The three basic objectives of VPAs (Abdullah 2011) are as follows.

1. Establish control and licensing systems providing government-backed guarantees of legality in the producing country.
2. Legality definition based on laws and procedures of the timber-producing country.
3. Legality assurance system that covers chain of custody, verification, licensing, and independent monitoring procedures.

The agreements are legally binding on both parties and require the exporting country to develop systems that verify the legality of their timber exports to the EU. The partners agree on the standards and assurance systems, and a FLEGT license is required for imports once the agreement is implemented. National legality standards and other requirements are defined through an open and inclusive political process in the partner country with support from the European Commission (Abdullah 2011).

As a second element, the EU created legislation to ban illegally produced wood products from the EU market, known as the EU Timber Regulation, which is binding and applicable to all EU member

[1] Details available at <www.fsc.org>

states. This legislation aims to prohibit the sale of illegally harvested timber, reduce illegal deforestation, assure consumers that the products they buy are legal and provide a level playing field for timber traders. FLEGT-licensed timber products (and CITES-certified timber) will be considered to have been legally harvested. Countries with VPAs can export FLEGT-licensed timber into the EU without operators (importers) needing to make further legality checks (Green Lane), and FLEGT-VPAs and the EU Timber Regulation are mutually reinforcing. By 2013, all the countries exporting timber to EU should have VPA. The legislation is based on the principle of due diligence, which requires importers to permit the traceability of timber and timber products by providing access to information about their sources and suppliers, including details of compliance with the applicable legislation and the country of harvest. Due diligence procedures will be offered by EU-accredited monitoring organizations, but the type of violation and related sanctions are determined by member states.

Details regarding verification of legal timber under VPAs and Certification of Sustainable Timber are given in Table 1.

**Table 1** Verification of legal timber under Voluntary Partnership Agreements and Certification of Sustainable Timber

| | *Legal timber* | *Sustainable timber* |
|---|---|---|
| Scope | P&C on social, environmental, and economic issues required by law | P&C on social, environmental, and economic issues required by law and forest sustainability |
| Application | Bilateral agreement (VPA); legally binding | Voluntary |
| Status | Governmental | Non-governmental |
| Supporting legislation | EU Timber Regulation | PPP in certain body |
| Assessment legislation | Third-party monitors | Certification body |
| Issuance of compliance document | Licensing authority for timber export party monitor | Certification body |
| Development process | Relevant authorities and stakeholders | Stakeholders |
| Source of timber | All forest land | Permanent reserved forest |
| Benefits of implementation | Entry into all EU countries, acceptance under PPP of certain countries (UK) | Acceptance under PPP |

P&C – principles and criteria; PPP – public–private partnership; VPA – voluntary partnership agreements

**Source** Freezailah (2011)

# Annex 3

## THE LACEY ACT

The Lacey Act is a law introduced by US Congressman John Fletcher Lacey and passed in 1900. Considered the first wildlife conservation law in the USA, the Lacey Act mandates that trade in animal products sourced in violation of any law in any country is illegal under the US federal law. The Lacey Act was amended in 2008 to expand its protection to a broader range of plants and plant products.

The amended Act makes it unlawful to trade in any plant taken, possessed, transported or sold in violation of federal or state laws of the USA, laws of a native American tribe or any foreign law that protects plants. It also makes it unlawful to falsify or submit falsified documents, accounts or records of any plant covered by the Act. Agricultural crops are excluded from the Lacey Act Amendment, as are certain other plants. In effect since May 2008, the Lacey Act is aimed at individuals and companies that handle plant products and makes it mandatory to provide import declarations with details of the country of harvest and the species name of the plants contained in the product. The enforcement of the declaration requirement is being phased in by product type. The Lacey Act emphasizes the principle of "due care". In order to fully comply with the Lacey Act, illegal wood must be eliminated from the supply chain by practicing due care. In the US legal system, due care is a flexible concept defined as the degree of care that a reasonably prudent person would exercise under the same or similar circumstances. The Act does not provide details of how due care can be shown; it is the responsibility of each company in the supply chain to exercise due care and understand the origin of its forest products. The expectations of the level of due care also differ based on the level of professionalism expected from an individual or organization.

Contrary to European legislation, the Lacey Act compliance is not limited to actions at the point of entry in the USA and remains with the product. Therefore, a violation of the Lacey Act can occur at almost any point in a forest product supply chain.

**Source** <www.fsc.org>

# Annex 4

## IMPORTANT ECO-LABELS AND PAPER-BASED PRODUCTS

| *Name of the eco-label* | *Characteristic* | *Type of wood eco-labelled* | *Website* |
|---|---|---|---|
| Blue Angel | The oldest environmental eco-label; initiated by the German Ministry of the Interior; it is now administered by the Federal | Wood- and paper-based products covered building materials, different types of paper and cardboard, and furniture Environmental Agency | www.blauer-engel.de |
| Bra Miljöval (Good | The eco-label from the Swedish Society for Environmental Choice) | Wood-based products covered include Nature Conservation, started in 1988 | www.snf.se/bmv/english.cfm various types of paper |
| Environmental Choice Program | Owned by the Canadian government and administered by Terra Choice Environmental Marketing | Wood- and paper-based materials covered include building raw materials, flooring, office furniture, and various types of paper | www. environemental choice. com |
| Eco Mark | Administered by the Japan Environment Association | Various types of paper, board wood, furniture, and packaging materials. | www.ecomark.jp/english/nintei.html |
| Environmental Choice | A voluntary, multiple specifications labelling programme endorsed by the New Zealand government and managed by the New Zealand Eco-labelling Trust | Wood-based products include various types of paper, furniture, and flooring products | www.enviro-choice.org.nz |
| EU Flower | Started in 1992 under the European Union Eco-labelling board; EU Flower is active throughout the European Union and in Norway, Liechtenstein, and Iceland | Wood-based products covered include various types of paper and building materials | www.ec.europa.eu/environment/ ecolabel/ index_en.htm |

*Contd...*

*Contd...*

| *Name of the eco-label* | *Characteristic* | *Type of wood eco-labelled* | *Website* |
|---|---|---|---|
| Green Seal | Developed by Green Seal Inc., an independent non-profit organization | Wood-based products covered include various types of paper, furniture, particle board and fibreboard, and food packaging materials. | www.greenseal.org/certification/ enviromental.cfm |
| Greenguard | Products certified meet requirements of the US Environmental Protection Agency, US Green Building Council, and Germany's Blue Angel eco-label | Buildings | www.greenguard.org |
| Good Environmental Choice | Designed by Good Environmental Choice Australia LTd | Wood- and paper-based products covered include various types of paper, flooring products, packaging materials, furniture, and recycled and reclaimed timber | www.aela.org.au/ standardsregister. htm |
| The Swan | The official Nordic eco-label introduced by the Nordic Council of Ministers | Certifies some paper products; it also certifies that durable wood products do not incorporate heavy metals or biocides and are produced from sustainably managed forests | www.svanen.nu/Eng/ |

**Source** <www.globalecolabelling.net>

There may be products bearing eco-labels that do not actually meet the label's environmental standards. The International Organization for Standardization (ISO) and other institutions provide guidance on general labelling standards to help in selecting eco-labels, some of which are as follows.

- ISO Standards 14020 through 14025 provide guidelines for eco-labels for first- and third-party verification.[1]
- The US Federal Trade Commission provides guidance on the use of eco-labels and the use of environmental marketing claims.[2]
- Consumer Reports Eco-labels provides guidance, scorecards, and comparisons of eco-labels in the USA.[3]
- The Global Eco-labelling Network provides background information, links to national members, and so on.[4]
- The UK government's Green Claims Code provides guidance on statements, symbols, descriptions, and verification.[5]

[1] Details available at <www.iso.org>

[2] Details available at <www.ftc.gov/bcp/grnrule/guides980427.htm>

[3] Details available at <www.greenerchoices.org/ecolablels/eco-homecfm>

[4] Details available at <www.gen.gr.jp/eco.html><

[5] Details available at <www.defra.gov.uk/environment/consumerprod/gcc/pdf/gcc.pdf>

# Annex 5

## COMPARISON OF FOREST CERTIFICATION SCHEMES OF INTEREST TO FOREST OWNERS OF THE USA (LAST REVISED 8 NOVEMBER 1999)

| *General feature* | *American Tree Farm System* | *Forest Stewardship Council* | | *Green Tag Forestry* | *International Organization for Standardization* | *Sustainable Forestry Initiative* |
|---|---|---|---|---|---|---|
| Sponsor | American Forest Foundation | Scientific Certification Systems | Smart Wood | National Forestry Association | International Organization for Standardization | American Forest and Paper Association |
| Scope | National | International | International | National | International | National |
| Director costs | Minimal | Mod. Exp. | Mod. Exp. | Moderate | Mod. Exp. | Mod. Exp. |
| Forest inspections (group certification) | √ | √ | √ | √ | √ | |
| Inspection term | | √ | √ | | √ | √ |
| Third-party verification | Unclear | √ | √ | Unclear | √ | Optional |
| Region-specific rules | In some areas | In some areas* | | Proposed | No | No |
| Status | Established | Established | Established | In development | Established | Established |
| Current rules and programmes specific to family forests | √ | √ | √ | √ | | √ |
| Chain of custody | | √ | √ | √ | | |
| Eco-label | | √ | √ | √ | | |

*Contd...*

*Contd...*

| *General feature* | *American Tree Farm System* | *Forest Stewardship Council* | | *Green Tag Forestry* | *International Organization for Standardization* | *Sustainable Forestry Initiative* |
|---|---|---|---|---|---|---|
| **General performance measures** | | | | | | |
| Compliance with existing laws | √ | √ | √ | √ | √ | √ |
| Requires written forest management plan | √ | √ | √ | √ | | |
| Public availability of certification summary | | √ | √ | | | |
| **Criteria focus (in addition to applicable laws)** | | | | | | |
| *Management and ecology* | | | | | | |
| Biodiversity (includes endangered species) | √ | √ | √ | √ | | √ |
| Chemical, pesticide, and herbicide use | | √ | √ | | | |
| Forest aesthetics | √ | | | | | √ |
| Forest health | √ | √ | | | | √ |
| Landscape | | √ | √ | | | |

*Contd...*

*Contd...*

| *General feature* | *American Tree Farm System* | *Forest Stewardship Council* | | *Green Tag Forestry* | *International Organization for Standardization* | *Sustainable Forestry Initiative* |
|---|---|---|---|---|---|---|
| Non-timber products (includes recreation) | √ | √ | √ | √ | | √ |
| Protect and enhance fish and wildlife habitat | √ | √ | √ | | | √ |
| Protection of special sites | √ | √ | √ | √ | | √ |
| Reforestation | √ | √ | √ | | | √ |
| Silvicultural treatments | | √ | √ | | | |
| Stream protection and water quality | | √ | √ | √ | | √ |
| Sustained yield | | √ | √ | | | √ |
| ***Operations*** | | | | | | |
| Road design, construction, and maintenance | | √<br>√ | √<br>√ | √ | | |
| Skidding and yarding | | | | | | |
| Slash disposal and product utilization | √ | √ | √ | √ | | |

*Contd...*

*Contd...*

| *General feature* | *American Tree Farm System* | *Forest Stewardship Council* | | *Green Tag Forestry* | *International Organization for Standardization* | *Sustainable Forestry Initiative* |
|---|---|---|---|---|---|---|
| Tree felling | | √ | √ | | | |
| *Social and economic* | | | | | | |
| Community relations | | √ | √ | √ | | √ |
| Documentation and monitoring (beyond plan) | | √ | √ | √ | √ | √ |
| Forestry contractor relations | √ | √ | √ | | | √ |
| Indigenous rights | | √ | √ | | | |
| Landowner tenure, rights, and responsibilities | | √ | √ | √ | | √ |
| Long-term economic viability | | √ | √ | √ | | √ |

√ Indicates specific reference to a particular criterion; it does not evaluate the cumulative effects of several criteria together.

* Regional guidelines may be available through either the Forest Stewardship Council or Smart Wood and may be different from that presented here.

**Sources**

1. American Tree Farm System Standards and Guidelines, March 1999
2. The Forest Conservation Program, October 1995
3. Smart Wood Certification Guidelines, April 1998
4. Green Tag Forestry Document, September 1999
5. SFI Voluntary Verification Indicators, 1999 edition

# Annex 6

## HOW MAJOR INTERNATIONAL CERTIFICATION SCHEMES ADDRESS SELECTED ASPECTS OF SUSTAINABLE FOREST MANAGEMENT

| *Aspects of sustainable forest management* | *Forest Stewardship Council* | *Programme for the Endorsement of Forest Certification* |
|---|---|---|
| Social issues | Four principles of the FSC system include various social concerns: tenure and user rights and responsibilities, indigenous people's rights, community relations, and workers' rights. Principles related to HCVF also address social aspects for areas of archaeological, historical or cultural value. Standard setting processes at the national and sub-national levels are conducted in a transparent way and involve all interested parties. | Requires compliance with ILO core conventions. PEOLG criteria and indicators address issues of occupational safety and health as well as accessibility to recreation and maintenance of sites with cultural or spiritual values. ATO/ITTO criteria and indicators for SFM require that legal and customary rights of local populations with respect to ownership, use, and tenure are clearly defined, acknowledged, and respected, as well as engagement with informed stakeholders (PEOLG, ATO/ITTO, Principles, Criteria, and Indicators (PCI) for SFM of African natural tropical forests). |
| Special places | Principle 9 addresses HCVFs, which are areas to be managed in such a way that these values are maintained or enhanced. HCVFs include: forests that contain globally, regionally or nationally significant concentrations of bio-diversity values; globally, regionally or nationally significant large landscape level forests; rare, threatened or endangered ecosystems; forest areas providing basic services of nature in critical situations; forest areas fundamental to meeting basic needs of local communities; forest areas critical to local communities'traditional cultural identity | Forest management should maintain or enhance biodiversity and protect soil and water. Sites of historical or spiritual significance should be respected and protected as specified by by international guidelines and standards.<br>Different requirements specified by international standards, criteria and indicators, and requirements for SFM, for instance PEOLG Criterion 4.2i—special key biotopes in the forest such as water sources, wetlands, rocky outcrops, and ravines should be protected or, where appropriate, restored when damaged by forest practices. |

*Contd...*

*Contd...*

| *Aspects of sustainable forest management* | *Forest Stewardship Council* | *Programme for the Endorsement of Forest Certification* |
|---|---|---|
| Forest plantations | Principles 6 and 10 of the FSC principles address forest plantations. Certified forest plantations should meet a set of requirements concerning the following.<br>• Representation on landscape<br>• Time of establishment<br>• Design of the management blocks (that is, blocks promote biodiversity)<br>Conversion of forests to plantations or non-forest land uses should not occur except in circumstances where conversion entails a very limited portion of the forest management unit, does not occur in high conservation value areas, and will deliver long-term conservation benefits. | Management standards for forest plantations are to be compliant with ITTO and PEOLG. |
| Chemicals | Principle 6 of FSC addresses chemicals. Chemicals should be minimized. Integrated pest management is a preferred approach, that is, to minimize chemical use through alternative prevention and biological control techniques. Documentation, monitoring, and control are required, and certain chemicals are banned. | Use of pesticides and herbicides should be minimized, used in controlled manner, and take into account appropriate silvicultural alternatives and other biological means. Compliance with PEOLG, ATO/ITTO criteria and indicators for SFM, as well as various ITTO guidelines for SFM |
| Clear-cuts | Principle 6 of FSC addresses clear-cuts. Restrictions on size and location vary among national/regional standards as long as ecological functions and values are maintained intact, enhanced or restored. | Management plans, including clear-cutting, should be based on legislation as well as existing land-use plans and adequately cover forest resources. Regeneration, tending, and harvesting should be carried out in time and manner that do not reduce the productive capacity of the site. |

*Contd...*

*Contd...*

| *Aspects of sustainable forest management* | *Forest Stewardship Council* | *Programme for the Endorsement of Forest Certification* |
|---|---|---|
| GMOs | Use of GMOs is prohibited; addressed in Principle 6 of FSC. | As required by PEOLG, native species and local provenances should be preferred where appropriate. Introduced species, provenances or varieties producing negative impacts on ecosystems and on the genetic integrity of native species and natural provenances should be avoided or minimized as should those not thoroughly evaluated. |

ATO – African Timber Organization; FSC – Forest Stewardship Council; GMO – genetically modified organism; HCVF – high conservation value forest; ILO – International Labour Organization; ITTO – International Tropical Timber Organization; PEOLG – Pan-European Operational Level Guidelines; SFM – sustainable forest management

**Source** <www.sustainableforestprods.org/node>

# Annex 7

## IMPLEMENTATION OF CERTIFICATION BY FOREST OWNERS

**Step 1 Evaluation of the need for forest certification**

The forest manager should be sure that demand for certification is real; that certification will confer a competitive advantage; and that the benefit from certification will outweigh its cost.

**Step 2 Undertake a preliminary analysis**

If a decision is taken to go ahead with certification, the owner should first prepare by undertaking a preliminary analysis. This involves the selection of standards and their local interpretation, followed by a baseline assessment of current activities to measure their adequacy against the selected standards. If possible, the preliminary analysis will also identify and prioritize the environmental and social effects of forestry operation. This will help to confirm the adequacy of objectives and targets during assessment.

**Step 3 Establish, implement, and evaluate an environmental management system**

Where inadequacies have been identified as part of Step 2, modifications will be required. These may be minor or major, perhaps even requiring a total rethink as to how the forestry activities are undertaken. In all cases, it is recommended that modifications are made within the context of an environmental management system (EMS). This allows for clear and feasible objectives and targets to be set; procedures to be defined to meet set targets; and a programme of internal monitoring to measure progress. Such an approach also provides an auditable framework for the assessor during certification and is likely to accelerate the certification process. More importantly, a strong EMS is focused on future improvement. It is more likely to provide an assurance that the certified forest management unit (FMU) will continue to meet the demands of the certification programme.

**Step 4 Invite certification of quality forestry practices**

The bulk of the certification process takes place once the operation is well prepared and success is likely to be achieved. However, certification is not an end-of-pipe, black-and-white, no-or-yes process. In reality, the forest manager will have started a process of dialogue and understanding with the certification body from the beginning.

In this way, award of the certificate is the culmination of a process resulting in improved environmental and social management of the forest in question and improved operating performance of the FMU. The certification report is not simply a statement of whether a particular forest operation has complied with the standards in question. Recommendations are made for further improvement and actions requested to further strengthen management in problem areas. Usually, time limits for the completion of specific activities or the attainment of set targets are agreed upon between the certification body and the forest manager. The above steps are illustrated below showing the main steps of the certification process, identifying the responsibilities of the certification body and those of the forest owner or operator.

# Annex 8

## COMPARING DIFFERENT FOREST CERTIFICATION WITH REGARD TO PROTECTION OF BIOLOGICAL DIVERSITY

| *Indicator* | *Forest Stewardship Council* | *Sustainable Forestry Initiative* |
|---|---|---|
| The status of forest-dependent species at risk of not maintaining viable breeding populations, as determined by legislation or scientific assessment | Safeguards shall exist that protect threatened, and endangered rare, species and their habitats. Conservation zones and protection areas shall be established. Ecological functions and values shall be maintained intact, enhanced, or restored, including genetic, species and ecosystem diversity. | Programme participants shall apply knowledge gained through research, science, technology, and field experience to manage wildlife habitats and contribute to the conservation of biological diversity. |
| Biological diversity | Landowners are required to maintain, enhance, or restore the long-term integrity of natural habitats, ecological processes, soil, water, and stand development. | Each company develops its own policies, programmes, and plans to contribute to the conservation of biological diversity and manage sites of ecological significance. |
| Stand age-class distribution | Forest owners and managers maintain or restore portions of the forest to the range and distribution of age classes of trees that result from processes that would naturally occur on the site. | Programme participants shall have policies, programmes, and plans to promote habitat diversity at the stand and landscape level. |
| Habitat diversity | A diversity of habitats for native species is protected, maintained, and enhanced. | Programme participants shall apply knowledge gained through research, science, technology, and field experience to manage wildlife habitats and contribute to the conservation of biological diversity. |

**Source** Washburn and Block (2001)

# Annex 9

## MAIN RECOMMENDATIONS OF THE NATIONAL FOREST CERTIFICATION COMMITTEE

The main recommendations of the Committee, as stated in the report, are as follows.

1. Setting up of an Indian Forest Certification Council, as a non-state multi-stakeholder independent entity.
2. The Government of India may consider providing corpus core funding, which could be around ₹100 million.
3. Meanwhile
   - A suitable organization may be assigned to host the Secretariat of the proposed IFCC at the earliest and intimate work for setting up of the National Forest Certification Scheme.
   - The NFCC Core Group, duly re-nominated, based on its expertise, may continue to facilitate the process of setting up IFCC and NFCS related work.

# Annex 10

## CONSULTATION WITH DIFFERENT STAKEHOLDERS

Although the National Forest Certification Committee (NFCC) had more than 40 members on board and submitted its report to the Ministry of Environment and Forests (MoEF), it has constituted a Core Committee to have one-to-one consultations with different stakeholders grouped into different categories with regard to their stakes. The different groups include certifying bodies, accreditation bodies, and non-governmental organizations involved in forest certification, other ministries at the Government of India level, state forest departments, paper and pulp industries, small farmers, and traders and exporters of value-added forest products. The core committee was chaired by DGF/ADG with other members. The composition of the Core Committee is enclosed as Annexure A. A set of questions has been prepared (Annexure B) to have a much more focused discussion on the structure and administrative set-up of the proposed Indian Forest Certification Council. The main outcomes of the one-to-one consultations with different groups are as follows.

- All the groups understand the urgency for setting up an independent Forest Certification Council irrespective of its name.
- Although the majority says that it should be an independent non-governmental body, the involvement of government agencies directly dealing with forest management and the trade, export, and import of timber and timber products cannot be overlooked. The officials from the concerned ministries have to be an integral part of the council along with other non-official members and academicians.
- There is a deep realization that due to the absence of a proper forest certification mechanism in our country, the entire business chain, right from raw material producers to product manufacturers, marketing chain, and small growers/farmers are making a loss. The handicrafts people, wood-carving communities, and exporters of value-added timber products are at present worst affected.
- Regarding the composition of the council, it has been opined by most of the groups that since the government owns almost 95% of the forest, the lead should be taken by the MoEF as a facilitator,

and it should take the initiative to set up a Forest Certification Council to facilitate all available types of forest certification equally. The MoEF should initially take the lead and then remain as a support for the council without being directly involved in its functioning. The details regarding the functioning of MoEF and the composition of Indian Forest Certification Council along with terms of references can be finalized in later discussions.

- It has been suggested by all groups that initial funding in the form of seed money/corpus fund should be given by the MoEF for the initial years. Over a period, the council should generate its own funding through its activities and should become a self-sustaining council.
- The proposed composition and TORs are enclosed.
- It has been emphasized by all groups that the council will facilitate both types of certification—chain of custody (CoC) and forest management unit (FMU)—in forests/plantations. There is an opinion that the whole certification process cannot be fragmented, and it needs to have both CoC and FMUs as major components.

### FMUs Certification in the Indian Context

- Regarding importance of FMU certification in India where export of round logs and swan timbers is prohibited as per the Exim Policy, there is a strong support for FMU certification as it exhibits that forests are managed as per best management practices by independent third-party bodies, which are recognized all over the world.

### Phased Approach to Forest Certification

- Regarding adopting a phased approach to forest certification, the consensus is that the phased approach should be adopted in both types of certification, that is, in CoC and FMUs. In CoC, the certification process needs to be prioritized and focus given to addressing the problems of small traders/exporters associated with value-added timber product exports.
- It was suggested that while adopting the phased approach, some of the forest areas can be taken up first for FMU certification as sample units either as per their bio-geographic zones or in a few divisions as per working plan prescriptions where harvesting

is suggested in the next couple of years. Even these certified forest areas will lead to better revenue from forests as well as better remuneration from private plantations who can take FMU certification.

- Initially, the certification process needs to focus on streamlining CoC certification with the initiative of the Government of India and also for an FMU certification for plantations (outside the forest) and to certify the raw materials available outside the forest, which are mainly utilized by small traders and exporters for handicrafts, toys, and carved items (group certification). The species mainly used for these units are shisham, mango, and acacia. A procedure needs to be put in place for these small entrepreneurs to get the raw materials (shisham and acacia) on a sustainable basis with the help of concerned state forest departments by earmarking a quantity on an annual basis from the forest depot.

### International Certification Standards

- With regard to utilizing the available international certification standards available in the market, there is a strong opinion that these standards can be utilized in the Indian context with a little modification for the initial couple of years. The council in the long run may develop national standards, which will take a few years to develop, but the choice of certification standards to be followed by users/owners of forests would totally be dependent on these users, that is, it will be market driven and based on the potential market premium that each certification standard/ logo gives the product. Whatever national certification systems exist in different countries, these are subsets of what international certification systems like the Forest Stewardship Council (FSC) and the Programme for the Endorsement of Forest Certification (PEFC) schemes have adopted.
- Regarding utilization of the criteria and indicators of sustainable forest management (SFM) developed by the Indian Institute of Forest Management (IIFM), Bhopal, as the Bhopal India Process (it is still not formally submitted by the IIFM to the ministry although almost finalized), there is a very strong apprehension that the Bhopal India Process can be only a part of the parameters/principles set up for forest certification. Since the whole exercise of the Bhopal

India Process was initiated almost a decade back and to assess the feasibility of SFM in different states (eight states), at best it can be taken up as a benchmark whenever the ministry/council initiates the development of national standards. Another angle that needs to be looked into is that these national standards need to be endorsed by an international agency as part of international processes that can be adhered to because the credibility to bring in a premium for any certified product under its standard is very important whenever we try to develop our own national standards for forest certification.

- The majority agrees that the issue of setting up standards by the council for its functioning depends on the mandate of the council. In case of a national certification scheme, the council can establish a standards development group by taking into account the various national certification schemes in practice.

In Malaysia, the Malaysian Timber Certification Council (MTCC), which has been created and registered as an independent organization, is the owner of the scheme developed. In other words, custodian MTCC is the owner of the Malaysian Timber Certification Standards.[1,2]

- Whenever standards are to be developed, many parameters such as the type of standards, cost of operation for certification users, potential to access the appropriate market and credibility need to be taken into account to make the certification programme a viable alternative to potential users.

## Accreditation and Certifying Bodies

- The stakeholders accept that certifying bodies shall have to be accredited by a reputed authority to make them more credible. It is always better that certification bodies should be accredited by the council and a national accreditation body like the National

[1] It must be decided whether the Indian Forest Certification Council shall be the scheme owner, on the lines of the Malaysian body, for the national forest certification system after developing its own standard or whether it will adopt the standards available in the market after modifying them to the Indian context. In case of voluntary standards, an independent body with balanced stakeholders (social, environmental, and economic) is appropriate. Being a stakeholder of forestry resources in the country, the MoEF and other relevant government institutions can be one of the representatives of the stakeholders.

[2] The World Wide Fund has formulated a working group with multi-stakeholders who work on developing standards mainly adopting FSC's international standards in the Indian context.

Accreditation Board for Certification Bodies (NABCB) or the Bureau of Indian Standards (BIS). These accreditation bodies will give accreditation to the certification bodies as per their general guidelines (ISO guide 64, 62) and additional requirements should be provided by the council.

- On the issue of setting of processes for accreditation boards, since the agencies available in India (NABCB or BIS) have the processes in place, the responsibility may be given to any of them. However, additional guidelines are to be provided by the council for forest certification. It has been advised by different groups that the existing accreditation boards already in business can be chosen given the expertise in domain and generic process. Otherwise, creating a new dedicated accreditation board would be a very costly and time-consuming exercise. Also it may be considered a non-independent or controlled process.
- Regarding the role of a standards-setting body and its identity, it has been very categorically emphasized that the standards-setting body has to be very clear in terms of process, that is, transparent, independent, and unbiased and should be financially independent. Regarding its identity, an independent body automatically will have more creditability for users, but the setting up of such a body has serious financial implications.

## Annex 10.A

Core Committee formed by Ministry of Environment & Forest on 3rd December, 2010 [O.M. No. 3-6/2001-SU (Vol. III)]

### CONSTITUTION OF CORE COMMITTEE

1. Director General of Forests & Special Secretary (Chairman)
2. Additional Director General (Forest), MoEF, New Delhi (Member)
3. Additional Director General (Wildlife), MoEF, New Delhi (Member)
4. IGF (EAP), MoEF, New Delhi (Member)
5. Director, Indian Institute of Forest Management of his representative (Member)

6. Director General, Indian Council of Forestry Research & Education of his representative (Member)
7. Deputy Inspector General of Forests, FP Division, MoEF, New Delhi (Member)
8. Deputy Inspector General of Forests, SU Division, MoEF, New Delhi (Member Secretary)

### Objectives

- Interact with different stakeholder on one to one basis
- Discuss various issues related to Forest Certification Mechanism need for Forest Certification Council in India, its composition, terms of reference, objectives and source of funding and also thrust of the proposed council on various stakeholders.
- Go through the recommendations of the report submitted by National Forest Certification Committee

### Methodology

- The Committee shall develop a common questionnaire on the various issues to have a better interaction with the stakeholders
- Records the discussion
- Report the major points/issues emerged out from the discussion

## Annex 10.B

### QUESTIONNAIRE FOR CERTIFICATION

1. What is the need to develop a Certification mechanism in India?
2. Who is getting the loss (in terms of value of the product) at present because of absence of a suitable mechanism available for Forest Certification in India?
3. What is the utility of the certification in the background of India?
4. Is certification process dealing with government?
5. What are the other certification mechanisms available globally?
6. Is it necessary that Certification Council should not be a part of government?

7. Do we require to set-up Indian Certification Council as an Independent Body?
8. If so what should be the mechanism for constituting/developing National Forest Certification Council?
9. Is it necessary that any Certification Council at national level has to be endorsed by FSE/PEFC?
10. Is there any agency in India working for the certification of forestry products?
11. In the Forestry Sector Certification is required only for NTFPs or for timber also?
12. Shall we go ahead with Product Certification (CoC) or also go for FMU certification along with CoC?
13. What is the need for FMU Certification in India where most of the forest areas are government owned?
14. Which one is more important Management Certification or Chain of Custody Certification?
15. What should be the role of government over Certification Council?
16. Whether IFCC will be cost effective and credible in International Market? What will be tools to make it more effective and credible?
17. Can we use standards setup by international bodies for Certification in India Context?
18. Do we need to develop a set of Standards for Forest Certification?
19. If so, before setting standards, is it necessary to setup rules for standards in accordance with the FSC or PEFC?
20. How Sustainable Forest Management is linked with Certification?
21. How far criteria and indicators developed at national level are useful in certification process?
22. Are these criteria and indicators developed feasible at the FMU level?
23. Do we have to develop separate set of C&I for NTFPs?
24. What is the role of the State Forest departments in the SFM and Certification process?

# Annex 11

## PROPOSED CONSTITUTION OF THE INDIAN FOREST CERTIFICATION COUNCIL

The Indian Forest Certification Council (IFCC) shall be registered under the Society Act. It will be an independent body, a not-for-profit organization formed with the aim of spreading, recognizing, and rewarding genuine sustainable forest management (SFM) initiatives in India and abroad.

The council shall consist of a governing body with an executive council to deal with the three aspects of forest certification, namely, standard setting and updating, certification and auditing, and promotion of the IFCC and sustainable forest product trade in India and abroad. There will be a body each for the three core areas of work, comprising members appointed by the executive council. Each body undertaking one of the core areas of work will have a group of support staff attached.

Almost 93% of the forest area in India is owned by the government. Forests are managed based on the sound principles of SFM through a system of working plans. The Joint Forest Management initiative of the Government of India has been overall successful in conservation, preservation, and sustainable benefit sharing from forests. A growing forest cover in India, as evident from the biennial assessment being conducted by the Forest Survey of India, Ministry of Environment and Forests (MoEF), through analysis of satellite data, corroborates this fact. The MoEF is the major stakeholder, and representatives of stakeholders in SFM will be represented at each level of the organizational structure of the proposed council.

**Name of the body:** Indian Forest Certification Council (IFCC)

**Field of operation**

- Forest management certification
- Chain of custody certification
- Plantation certification
- Group certification for small forests/agro-forests/trees outside forests
- Northwest Forest Plan (NWFP) and MAP certification

**Area of operation:** Indian Subcontinent

**Objectives of IFCC**

- To certify forest management units, plantation areas, forest-based industries, small and marginal forest owners as a group who manage their forests responsibly complying with the set standards.
- Set and continuously improve national set of standards for forest management, chain of custody, plantation and group certification, and NWFP suiting the changing needs of different stakeholders.
- Promote national and international trade of sustainably sourced forest products.
- Promote the Indian Forest Certification Scheme (IFCS) in India and abroad to build linkages with environment-sensitive consumers/ traders/nation.
- To develop and expand its network through partnership with other similar and relevant agencies and organizations.
- Provide recommendations, advisory and other support services to the timber and NWFP trade and industry.
- To disseminate information related to forest councils, SFM, timber trade, NWFP, international market information that benefits one and all by periodically publishing magazines, journals, and guidelines.

**Members:** Governing body/general body—10 members excluding one chairman and one vice-chairman

**Executive council:** Total 20 members, 10 members from the general body and 10 nominated by the general body. This includes CEO, one Executive Council member, and one deputy CEO.

**Technical body for updating standards:** Total six members—three members (nominated) from the executive council and three nominated by the governing body.

**Body for certification and auditing:** Total six members—three members (nominated) from the executive council and three nominated by the governing body.

**Body for promotion of SFM and forest council:** Total six members—three members (nominated) from the executive council and three nominated by the governing body.

# Bibliography


Abdullah, A. L. 2011. Personal discussion on 31/3/2011 at FLEGT, Malaysian office, Kuala Lumpur

Adams, D. A. 1993. *Renewable Resource Policy: the legal-institutional foundations*. Island Press: Washington, DC

Anderson, R. C., and E. N., Hansen. 2004. Determining consumer preferences for eco-label forest products: an experimental approach. *Journal of Forestry* 102(4): 28–32

Atyi, R., and M., Simula. 2002. Forest certification: pending challenges for tropical timber. *ITTO Technical Series No 19*. Yokohama: International Tropical Timber Organization

Azevedo-Ramos, C., O., de Carvalho, Jr., and R., Nasi. 2002. *Animal indicators: a tool to assess biotic integrity after logging in tropical forests*? Jakarta: Center for International Forestry Research

Bass, S. 1998. Introducing forest certification—a report prepared by the Forest Certification Advisory Group (FCAG) for DGVIII of the European Commission. European Forest Institute, Discussion Paper 1. Details available at <www.efi.int/files/attachments/dp_01.pdf>

Bass, S., and M., Simula. 1999. Independent certification/verification of forest management. Background paper for the World Bank/WWF Alliance Workshop, 9–10 November 1999, Washington, DC. Details available at <www.giz.de/Themen/de/dokumente/en-d28-inenpenennt-certification---verification-forest-manag.pdf>

Bass, S., K., Thornber, M., Markopoulos, S., Roberts, and M., Grieg-Gran. 2001. Certification's impacts on forests, stakeholders, and supply chains. *Instruments for Sustainable Private Sector Forestry Series*. London: International Institute for Environment and Development

Bettinger, B., K., Boston, J. K., Siry, and D. L., Grebner. 2009. *Forest Management and Planning*, 1st edn. Academic Press

Carey, C. 2008. E047 *Governmental Use of Voluntary Standards Case Study 2: Bolivia and FSC standards*. London: ISEAL Alliance

Cashore, B., and C. L., McDermott. 2004. *Global Environmental Forest Policy: Canada as a constant case comparison of select forest practice regulations*. Victoria, BC: International Forest Resources

Castañeda, F. 2000. Criteria and indicators for sustainable forest management: international processes, current status, and the way ahead. *Unasylva* 51: 34–40

Castañeda, F., C., Palmberg-Lerche, and P., Vuorinen. 2001. Criteria and indicators for sustainable forest management: a compendium. *Forest Management Working Paper FM/5*. Rome: Food and Agriculture Organization. Details available at <www.fao.org/DOCREP/004/AC135E/ac135e08.htm>

Cerda, A. 2000. Visión General del Mercado de la Certificación Forestal, XV Silvotecna, Concepción, Chile. Details available at <www.redmaderamuebles.cl/usuarios/corpa/doc/200506081343160.certificacion(CMPC).pdf>

Cerda, A., and V., Lira. 2002. The economics of sustainable forest management certification. Paper presented at the *Second World Congress of Environmental and Resources Economists,* Monterrey, California, 24–27 June 2002, organized by the Giannini Foundation of Agricultural and Resource Economics, University of California, Berkeley and Davis, and the Donald Bren School of Environmental Science and Management, University of California, Santa Barbara

Chesnutt, J., M., Joseph Cox, and W., Boyette. 2003. Forest certification costs and benefits: the experience of the Southern Center for Sustainable Forests. In *Proceedings, 2002 Society of American Foresters Annual Convention*. Bethesda, MD: Society of American Foresters, pp. 236–242

Clarke, J. 2000. Social and environmental aspects of the forest management certification process: a discussion of social assessment components in South Africa. *Instruments for Sustainable Private Sector Forestry, South Africa Series*. London and Pretoria: International Institute for Environment and Development and CSIR-Environmentek. Details available at <www.researchspace.csir.co.za/dspace/bitstream/10204/2485/1/Clarke_2000.pdf>

Cleveland, C. J., I., Kubiszewski, M., Miller, United Nations, P., Saundry. 2007. United Nations Conference on Environment and Development (UNCED), Rio de Janeiro, Brazil. In *Encyclopedia of Earth,* edited by

C. J. Cleveland (ed.). Washington, DC: Environmental Information Coalition, National Council for Science and the Environment

Cochrane, M. A. 2001. In the line of fires: understanding the plan. *Tropical Forest Update* 12: 10–11

Conroy, M. E. 2007. *Branded! How the "certification revolution" is transforming global corporations.* Gabriola Island, BC: New Society Publishers

Corvalan, C., S., Hales, A., McMichael, C., Butler, D., Campbell-Lendrum, U., Confalonieri, K., Leitner, N., Lewis, J., Patz, K., Polson, J., Scheraga, A., Woodward, M., Younes, and many MA authors. 2005. *Ecosystems and Human Well-being: health synthesis—a report of the Millennium Ecosystem Assessment.* Geneva: WHO Press

Costa, P. M., J., Salmi, M., Simula, and C., Wilson. 1999. Financial Mechanism for Sustainable Forestry: Working Draft. New York: United Nations Development Programme

Cubbage, F., D., Richter, M., Thompson, S. E., Moore, J., Edeburn, M., Chesnutt, J., Cox, and W., Boyette. 2003. Forest certification costs and benefits: the experience of the Southern Center for Sustainable Forests. In *Proceedings, 2002 Society of American Foresters Annual Convention.* Bethesda, MD: Society of American Foresters, pp. 236–242

Danielsen, F., D., Burgess, A., Balmford, P. F., Donald, M., Funder, J. O. G., Jones, P., Alviola, D. S., Balete, T., Blomley, J., Brashares, B., Child, M., Enghoff, J., Fjeldsa, S., Holt, H., Hubertz, A. E., Jensen, P. M., Jensen, J., Massao, M. M., Menoza, Y., Ngaga, M. K., Poulsen, R., Rueda, M., Sam, T., Skielboe, G., Sturat-Hill, E., Topp-Jørgensen, and D., Yonten. 2008. Local participation in natural resource monitoring: a characterization of approaches. *Conservation Biology* 23: 31–42

Danielsen, F., M. M., Mendoza, A., Tagtag, P. A., Alviola, D. S., Balete, A. E., Jensen, M., Enghoff, and M. K., Poulsen. 2007. Increasing conservation management action involving local people in natural resource monitoring. *Ambio* 7: 566–570

de Azevedo, T. R., and A. G., de Freitas. 2003. Annex 1: forest certification in Brazil—the parallel evolution of community forest management in the Brazilian Amazon and FSC certification. Details available at <www.forest-trends.org/documents/files/doc_321.pdf>

de Iongh, H., and G., Persoon. 2010. Monitoring the impact of certification. *ETFRN News* 51: 48–50. Details available at <www.etfrn.org/etfrn/newsletter/news51/Chapters/2.5Iongh-Persoon.pdf>

Dudley, N., J., Jeanrenaud, and F., Sullivan. 1995. *Bad Harvest? The Timber Trade and Degradation of the World's Forests*. London: Earthscan

Durst, P. B., P. J., McKenzie, C. L., Brown, and S., Appanah. 2006. Challenges facing certification and eco-labeling of forest products in developing countries. *International Forestry Review* 8(2): 193–200

Dwivedi, A. P. 1980. *Forestry in India*. Dehradun: Jugal Kishore and Company

ECE (Economic Commission for Europe). 1997. *Forest and Forest Industries Country Fact Sheets: Geneva timber and study papers*. New York and Geneva: United Nations

ETFRN (European Tropical Forest Research Network). 2010. Biodiversity conservation in certified Forests. *ETFRN News* 51: 1–228

Eyre, C., and J., Mundy. 1999. *Risk mitigation in forestry: linkages with Kyoto and sustainable forestry management—a review*. Washington, DC: Forest Trends

FAO (Food and Agriculture Organization). 1996. Forest products markets in 1996 and prospects for 1997. *Timber Bulletin* XLIX(6): 15–17

FAO (Food and Agriculture Organization). 2000. Report of the FAO/UNEP/ITTO/IIFM/USFS workshop on the development of national-level criteria and indicators for the sustainable management of dry forests in Asia. Bhopal, India, 30 November–3 December 1999, FAO Regional Office for Asia and the Pacific, Bangkok, Thailand

FAO (Food and Agriculture Organization). 2009a. *Proceeding of XIII World Forestry Congress*. Rome: FAO

FAO (Food and Agriculture Organization). 2009b. *State of the World's Forests 2009*. Rome: FAO

FAO/ECE (Food and Agriculture Organization/Economic Commission for Europe). 1996. Geneva Timber and Forest Study Papers. European Timber Trends and Prospects: into the 21st Century. Details available at <www.unece.org/fileadmin/DAM/timber/docs/sp/SP-11.pdf>

Fernholz, K., J., Howe, S., Bratkovich, and J., Bowyer. 2010. *Forest Certification: a status report*. Minneapolis, MN: Dovetail Partners Inc.

Fischer, C., F., Aguilar, P., Jawahar, and R., Sedjo. 2005. *Forest certification: toward common standards? Discussion Paper 5-10*. Washington, DC: Resources for the future. Details available at <http://rff.org/RFF/Documents/RFF-DP-05-10.pdf>

Forest Peoples Programme. 2006. *Protecting Indigenous Rights in the Republic of Congo through the Application of FSC Standards in Forest Plans: a review of progress made by Congolaise Industrielle des Bois (CIB) against FSC Principles 2 and 3*. Moreton-in-Marsh, UK: Forest Peoples Programme

Freezailah. 2011. Personal discussion on 31/3/2011 at FLEGT, Malaysian office, Kuala Lumpur

FSC (Forest Stewardship Council). 2009a. Comparative analysis between the FSC controlled wood requirements and PEFC, PEFC Germany, and SFI. Details available at <www.fsc-schweiz.ch/de/download-center/doc.../62-cwpefcbericht>

FSC (Forest Stewardship Council). 2009b. FSC impacts and outcomes: extracts from FSC literature review 2009. Details available at <http://pre.fsc.org/fileadmin/web-data/public/document_center/publications/FSC_Policy_Series/Impacts_lit_review_summary_EN.PDF>

Gan, J. 2005. Forest certification costs and global forest product markets and trade: a general equilibrium analysis. *Canadian Journal of Forestry* 35: 1731–1743

Ghazoul, J., and A., Hellie. 2000. Setting limits to ecological indicators of sustainable tropical forestry. *International Forestry Review* 2: 243–253

Granholm, H., T., Vahanen, and S., Sahlberg. 1996. *Background Document: intergovernmental seminar on criteria and indicators for sustainable forest management*. Helsinki, Finland: Ministry of Agriculture and Forestry

Gullison, R. E. 2003. Does forest certification conserve biodiversity. *Oryx* 37(2): 153–157

Handford, P., and R., Nussbaum. 2001. An introduction to forest certification. Details available at <www.ogm.gov.tr/yukle/orman_serf.pdf>

Hansen, E. 1997. Forest certification and its role in marketing strategy. *Forest Products Journal* 47(3): 16–22

Hansen, E., R., Fletcher, B., Cashore, and C., McDermott. 2006. Forest Certfication in North America. Details available at <http://extension.oregonstate.edu/catalog/pdf/ec/ec1518-e.pdf>

Hansen, E., K., Forsyth, and H., Juslin. 2000. Forest certification updates for the ECE region, summer 2000. Geneva Timber and Forest Discussion Papers. Details available at <www.unece.org/fileadmin/DAM/timber/docs/dp/dp-20.pdf>

Hardin, G. 1968. The tragedy of the commons. *Science* 162(1968): 1243–1248

Hayward, J., and I., Vertinsky. 1999. What managers and owners think of certification? *Journal of Forestry* 97(2): 13–17

Heske, F. 1938. *German Forestry*. New Haven: Yale University Press

Higman, S., J., Mayers, S., Bass, N., Judd, and R., Nussbaum. 2006. *The Sustainable Forestry Handbook*, 2nd edn. London: Earthscan

Hirschberger, P. 2005. The effects of FSC certification in Estonia, Germany, Latvia, Russia, Sweden, and the UK: an analysis of corrective action requests. Details available at <www.panda.org/downloads/.../fscsummaryanalysisallcountries.pdf>

Howard, A., R., Rice, and R., Gullison. 1996. Simulated financial returns and selected environmental impacts from four alternative silvicultural prescriptions applied in the neotropics: a case study of the Chimanes Forest, Bolivia. *Forest Ecology and Management* 89: 43–57

Hughell, D., and R., Butterfield. 2008. Impact of FSC certification on deforestation and the incidence of wildfires in the Maya Biosphere Reserve. Details available at <www.rainforest-alliance.org/sites/default/files/publication/pdf/peten_study.pdf>

IIFM (Indian Institute of Forest Management). 2000. *Bhopal India Process for Sustainable Management of Indian Forests*. Bhopal: IIFM

Irvine, D. 1999. Certification and community forestry: current trends, challenges, and potential—a background paper for the World Bank/WWF Alliance Workshop on Independent Certification. Details available at <www.giz.de/Themen/de/dokumente/en-d26-certification-and-community-forestry-trends-challenge.pdf>

ITTO (International Tropical Timber Organization). 2002. Challenges of growing certification. *Tropical Forest Update* 12(3): 1–3

ITTO (International Tropical Timber Organization). 2008. Developing forest certification: towards increasing the comparability and acceptance of forest certification systems worldwide. ITTO Technical Series 29. Details available at <www.pefc.org/images/stories/documents/external/enPDF.pdf>

ITTO (International Tropical Timber Organization). 2011. *Tropical Forest Update* 20(3)

Johns, A. D. 1987. The use of primary and selectively logged rainforest by Malaysian hornbills (Bucerotidae) and implications for their conservation. *Biological Conservation* 40: 179–190

Johns, A. G. 1996. Bird population persistence in Sabahan logging concessions. *Biological Conservation* 75: 3–10

Kanowski, P., D., Sinclair, and B., Freeman. 1999. International approaches to forest management certification and labelling of forest products: a review. Commonwealth of Australia

Kessler, M. 1999. Plant species richness and endemism during natural landslide succession in per-humid montane forest in the Bolivian Amazon. *Ecotropicas* 5: 123–136

Kiekens, J. 2003. Forest certification in North America: selected developments. 12th World Forestry Congress, Canada

Koch, P. 1992. Wood versus non-wood materials in residential construction: some energy related global implications. *Forest Products Journal* 42(5): 31–42

Landres, P. B., J., Verner, and J. W., Thomas. 1988. Ecological uses of vertebrate indicator species: a critique. *Conservation Biology* 2: 316–328

Layton, P. A., S. T., Guynn, and D. C., Guynn. 2002. Wildlife and biodiversity metrics in forest certification systems. Technical Bulletin No. 0857. Research Triangle Park, NC: National Council for Air and Stream Improvement, Inc

Lawson, S., and L., McFaul. 2010. *Illegal Logging and Related Trade: indicators of the global response*. London: Chatham House

Leslie, A. 2006. Out on a limb: the SFM conundrum. *Tropical Forest Update* 16(3): 31–32

Loehle, C., T. B., Wigley, S., Rutzmoser, J. A., Gerwin, P. D., Keyser, R. A., Lancia, C. J., Reynolds, R. E., Thill, R., Weih, D., White, Jr., and P. B., Wood. 2005. Managed forest landscape structure and avian species richness in the southeastern US. *Forest Ecology and Management* 214(1–3): 279–293

Lutteken, A., and K., Hagedorn. 1999. Concepts and issues of sustainability in countries in transition: an institutional concept of sustainability as a basis for the network. Details available at <www.fao.org/docrep/006/AD238E/ad238e08.htm>

Mannan, S., K., Kitayama, Y. F., Lee, A., Chung, A., Radin, and P., Lagan. 2008. Deramakot forest shows positive conservation impacts of RIL. *Tropical Forest Update* 18(2): 7–9

Markopoulos, M. D. 1998. *The Impacts of Certification on Community Forest Enterprises: a case study of the Lomerio Community Forest Management Project, Bolivia.* Oxford: Oxford Forestry Institute

Markopoulos, M. D. 1999a. *Community Forest Enterprise and Certification in Mexico: a review of experience with special reference to the Union of Zapotec and Chinantec Forestry Communities.* Oxford: Oxford Forestry Institute

Markopoulos, M. D. 1999b. *The Impacts of Certification on Campesino Forestry Groups in Northern Honduras.* Oxford: Oxford Forestry Institute

Martens, P., J., Rotmans, and D., de Groot. 2003. Biodiversity: luxury or necessity? *Global Environmental Change* 13: 75–81

Massao, M., M., Menoza, Y., Ngaga, M. K., Poulsen, R., Rueda, M., Sam, T., Skielboe, G., Stuart-Hill, E., Topp-Jørgensen, and D., Yonten. 2008. Local participation in natural resource monitoring: a characterization of approaches. *Conservation Biology* 23: 31–42

Mater, C. M. 1999. *Understanding Forest Certification: answers to key questions.* Washington, DC: Pinchot Institute for Conservation

MEA (Millennium Ecosystem Assessment). 2005. *Ecosystems and Human Well-being: biodiversity synthesis.* Washington, DC: World Resources Institute

Mechel, F., N., Meyer-Ohlendorf, P., Sprang, and R. G., Tarasofsky. 2006. *Public Procurement and Forest Certification: assessing the implications for policy, law, and international trade—comparing major certification schemes: FSC, PEFC, MTCC, and SFI.* Berlin: Ecologic—Institute for International and European Environmental Policy

MoEF (Ministry of Environment and Forests). 2006. *National Status Report on Forests and Forestry in India.* New Delhi: Survey and Utilization Division, MoEF

MoEF (Ministry of Environment and Forests). 2007. *Minutes of the First Meeting of Committee of Accreditation Criteria and Processes towards Forest Certification for Timber and Non-timber Forest Products.* New Delhi: Survey and Utilization Division, MoEF

Molnar, A., R., Butterfield, F., Chapela, P., Fuge, A. G., de Freitas, J., Hayward, J. -W., Jansens, M., Jenkins, S., Madrid, A., Martin, T. R., de Azevedo, M. Ridder, P., Smith, C., Soza, and A., White. 2003. *Forest certification and communities: looking forward to the next decade.* Washington, DC: Forest Trends

Moura Costa, P., J., Salmi, M., Simula, and C., Wilson. 1999. *Financial Mechanism for Sustainable Forestry.* New York: United Nations Development Programme

Nasi, R., P., Koponen, J., Poulsen, M., Buitenzorgy, and W., Rusmantoro. 2008. Impact of landscape and corridor design on primates in a large-scale industrial tropical plantation landscape. *Biodiversity and Conservation* 17: 1105–1126

Newsom, D., and D., Hewitt. 2005. The global impacts of smartwood certification: final report of the TREES Program for the Rainforest Alliance. Details available at <www.iatp.org/files/Global_Impacts_of_SmartWood_Certification_The.pdf>

Nussbaum, R., and M., Simula. 2004. Forest Certification: a review of impacts and assessment frameworks. Details available at <www.proforest.net/objects/publications/tfd-forest-certification-executive-summary.pdf>

NWF/NRCM/EA (National Wildlife Federation/Natural Resources Council of Maine/ Environmental Advocates). 2001. *A Comparison of the American Forest and Paper Association's Sustainable Forestry Initiative and the Forest Stewardship Council's Certification System.* Washington, DC: National Wildlife Federation

Oliver, R. 2010. Forest certification challenged by climate change and illegal logging concerns: certified forest products markets, 2008–2009. In *UNECE/FAO Forest Products Annual Market Review, 2008–2009.* Rome: FAO

Pearce, D., and D., Moran. 1994. *The Economic Value of Biodiversity*. London: Earthscan

Perera, P., and R. P., Vlosky. 2006. A history of forest certification. In *Louisiana Forest Products Development Center Working Paper No. 71*. Louisiana State University: Baton Rouge, Los Angeles

Pradère, A. 1989. *French Furniture Makers: the art of the Ebéniste from Louis XIV to the Revolution*. London: Philip Wilson Publishers Ltd

Prasad, R. 1999. Vision of model forests through sustainable forest management and community participation. *Tech Review IIFM Newsletter* II(1)

Princen, T., and M., Finger. 1994. *Environmental NGOs in World Politics: linking the local and the global*. Routledge: London

Rametsteiner, E., and P., Schwarzbauer. 1999. Introduction: the European markets for certified forest products. *Potential Markets for Certified Forest Products in Europe: EU-project FAIR-CT95-766*, edited by B. Pajari, T. Peck, and E. Rametsteiner. Joensuu, Finland: European Forest Institute. Details available at <www.efi.int/files/attachments/publications/proc25_net.pdf>

Rametsteiner, E., and M., Simula. 2001. Background paper for workshop on forest certification: forging novel incentives for environment and sustainable forest management. In *Forest Certification: forging novel incentives for environment and sustainable forest management*, edited by M. Simula, E. Rametsteiner, A. Blåsten, T. Green, and B. Pajari. Joensuu, Finland: European Forest Institute. Details available at <www.efi.int/files/attachments/publications/proc43_net.pdf>

Rangrajan, L. N. (trans). 1987. *The Arthashastra (Penguin Classics)*. New Delhi: Penguin Books India

Rao, V. S., K. P., Sagaraiya, C. A. R., Bhadran, and P., Venkatramany (eds). 1961. *Hundred Years of Indian Forestry*. Dehradun: Forest Research Institute

Rawat, A. S. 1991. *History of Forestry in India*. Delhi: Indus Publishing Company

Rawat, T. S., B. L., Menaria, D., Dugaya, and P. C., Kotwal. 2008. Sustainable forest management in India. *Current Science* 94(8): 996–1002

Roper, J. 1999. Deforestation: tropical forests in decline. In *Forestry Issues Paper*. Ottawa: Canadian International Development Agency.

Rudel, T., and J., Roper. 1996. Regional patterns and historical trends in tropical deforestation: a qualitative comparative and NGO clients on the conservation and management of analysis. *Ambio* 25: 160–166

Sachin, R. 2010. A visionary approach for the paper manufacturers in India. *IPPTA Journal* 22(1)

SAF (Society of American Foresters). 1999. Task Force on Forest Management certification Programs, 1999 Report. Details available at <www.safnet.org/policy/fmcp1999.html>

Sample, V. A. 2004. Sustainability in forestry: origins, evolution, and prospects. Pinchot Institute for Conservation, Washington, DC. Details available at <www.pinchot.org/?module=uploads&func=download&fileId=4>

Sheil, D. 2001. Conservation and biodiversity monitoring in the tropics: realities, priorities, and distractions. *Conservation Biology* 15: 1179–1182

Sheil, D., R., Nasi, and B., Johnson. 2004. Ecological criteria and indicators for tropical forest landscapes: challenges in search for progress. *Ecology and Society* 9(1): 7

Sheil, D., F. E., Putz, and R. J., Zagt (eds.). 2010. *Biodiversity Conservation in Certified Forests*. Wageningen, the Netherlands: Tropenbos International

Simula, M. 1999. *Certification of Forest Management and Labelling of Forest Products: discussion note on main issues*. Washington, DC: World Bank

Simula, M., E., Rametsteiner, A., Blåsten, T., Green, and B., Pajari (eds). 2001. Forest certification: forging novel incentives for the environment and sustainable forest management. EFI Proceedings No. 43, 2001. Joensuu: European Forest Institute. Details available at <www.efi.int/files/attachments/publications/proc43_net.pdf>

Slik, J. W. F. 2001. Macaranga and Mallotus (Euphorbiaceae) as indicators for disturbance in the lowland dipterocarp forests of East Kalimantan, Indonesia. *Tropenbos-Kalimantan Series 4*. Wageningen: The Tropenbos Foundation. [PhD dissertation]

Soza, C. 2003. Forest Certification in Guatemala: the certification process in the Maya biosphere reserve in Petén. In *Forest Certification and Communities: looking toward the next decade*, edited by A. Molnar. Forest Trends: Washington, DC

Spears, J. 2000. *Global Vision Research Project: summary of main findings emerging from phase 1 and their implications for further research*. Washington, DC: World Bank-WWF Alliance

Teng, C. L., H., Singh, and K. Y., Teng. 2009. *The first ten years (1999–2009): ensuring sustainable forest management*. Kuala Lumpur: Malaysian Timber Certification Council

Thornber, K. 1999. Overview of Global Trends in FSC Certificates. Instruments for sustainable private sector forestry. Forestry and Land Use Programme. Details available at <http://pubs.iied.org/pdfs/G00411.pdf>

Tikina, A., R., Kozak, and B., Larson. 2008. What factors influence obtaining forest certification in the U.S. Pacific Northwest? *Forest Policy and Economics* 10(4): 240–247

UN (United Nations). 1987. *Report of the World Commission on Environment and Development: our common future*. Geneva: UN. Details available at <http://conspect.nl/pdf/Our_Common_Future-Brundtland_Report_1987.pdf>

UNECE–FAO (United Nations Economic Commission for Europe–Food and Agriculture Organization). 1996a. Forest products markets in 1996 and prospects for 1997. *Timber Bulletin* 49(6): 32

UNECE–FAO (United Nations Economic Commission for Europe–Food and Agriculture Organization). 1996b. European timber trends and prospects: into the 21st century. In *Geneva Timber and Forest Study Papers*. New York/Geneva: United Nations

UNECE–FAO (United Nations Economic Commission for Europe–Food and Agriculture Organization). 1997. Forest and forest industries: country fact sheets. In *Geneva Timber and Forest Study Papers*. New York/Geneva: United Nations

Upton, C., and S., Bass. 1995. *The Forest Certification Handbook*. London: Earthscan

v. Alvensleben. 1992. Gründe für den Kauf von Lebensmitteln aus ökologischem Anbau. agungsreihe Neubewertung von Lebensmittelqualität. Hrsg: Katalyse/Buntstift Köln, Germany

van Dam, C. 2003. The economics of forest certification: sustainable development for whom? Paper presented at *The Latin American Congress on Development and Environment: local challenges of globalization*, Quito, Ecuador, 11–12 April 2003, organized by FLACSO

van Dam, J., and H., Savenije. 2011. *Enhancing the Trade of Legally Produced Timber: a guide to initiatives*. Wageningen, the Netherlands: Tropenbos International

van der Hoeven, C. A. 2007. The missing link: bridging the gap between science and conservation. PhD thesis, Department of Environmental Sciences, Resource Ecology Group, Wageningen University, the Netherlands, 152 pp.

van der Hoeven, C. A., H. H., de Iongh, V., Nijman, and B., Balen. 2000. *Biodiversity in disturbed ecosystems: a literature review of the use of fauna indicators for the assessment and monitoring of the levels of human disturbance in Bornean tropical lowland forests*. Wageningen, the Netherlands: NWO and the Tropenbos Foundation

van Kuijk, M., F. E., Putz, and R., Zagt. 2009. *Effects of Forest Certification on Biodiversity*. Wageningen, the Netherlands: Tropenos International. Details available at <www.pefc.org/images/stories/documents/external/forest_certification_and_biodiversity.pdf>

van Soest, D. 1998. *Tropical Deforestation: an economic perspective*. Capelle, the Netherlands: Labyrint Publication

Vidal, N., R., Kozak, and D., Cohen. 2005. Chain of custody verification: an assessment of the North American solid wood sector. *Forest Policy and Economics* 7(3): 345–355

Vincent, J., and C., Binkley. 1993. Multiple-use forestry. *Journal of Land Economics* 69(4): 370–376

Vlosky, R. P., L. K., Ozanne, and R. J., Fontenot. 1999. A conceptual model of US consumer willingness-to-pay for environmentally certified wood products. *Journal of Consumer Marketing* 16(2): 122–140

Vogt, K. A., B. C., Larson, J., Gordon, D. J., Vogt, and A., Fanzeres (eds). 2000. *Forest Certification: roots, issues, challenges, and benefits*. Boca Raton, Florida: CRC Press LLC. 400 pp.

Waggener, T. R. 1977. Community stability as a forest management objective. *Journal of Forestry* 75(11): 710–714

Washburn, M. P., and N. E., Block. 2001. *Comparing Forest Management Certification Systems and the Montreal Process Criteria and Indicators*. Washington, DC: Pinchot Institute

Wibowo, D. 2002. The challenges of growing certification. *ITTO Tropical Forest Update* 12(3). Details available at <www.itto.int/direct/topics/topics_pdf.../topics_id=1750000&no=1>

Yadav, M., P. C., Kotwal, and M., Menaria. 2008. *Forest Certification a Tool for Sustainable Management*. Bhopal: Indian Institute of Forest Management

# Glossary

**Accreditation authority** An independent third party that examines the organizational structure, responsibilities, practices, procedures, processes, and resources of a certification body in the pursuit of certification activities.

**Accreditation** A process where an authoritative body formally recognizes an auditing body as capable of carrying out the tasks necessary to certify a company's operation against a particular standard. It is also commonly referred to as certifying the certifiers.

**Accreditation schedule** A document issued by the accreditation authority, which details the areas in which a certification body has been assessed as competent to operate within.

**American Tree Farm System (ATFS)** A third-party certification standard focusing on the management practices of private landowners in the USA. To qualify for the ATFS certification, the landowner must own less than 10,000 acres of contiguous forestland and not be affiliated with any forest products' manufacturing facility.

**Applicant** (for certification) A person or body that seeks to obtain a license from a certification body.

**Assessment** A third-party examination and evaluation of a local forest management unit's systems to determine the degree of conformity against a specified standard and also to validate its effective implementation.

**Assessor** A person qualified and authorized to perform all or any portion of a certification assessment.

**Assessor's handbook** A document containing guidance and information to assist an assessor in the performance of assessment activities.

**Auditor (or certifier)** An accredited body qualified to inspect and verify that a company's practices conform to the certification standards.

**Best forest management practices** Equivalent to quality forestry activities and recognized as the best currently achieved performance for a particular forest type by a local forest management unit.

**Biodiversity** A measure of the richness of life forms, often divided into ecosystem, species, and genetic diversity.

**Biological control agents** Living organisms used to eliminate or regulate the population of other living organisms.

**Biological diversity values** Intrinsic, ecological, genetic, social, economic, scientific, educational, cultural, recreational, and aesthetic values of biological diversity and its components.

**Biological diversity** Variability among living organisms from all sources, including, inter alia, terrestrial, marine, and other aquatic ecosystems and the ecological complexes of which they are a part; this includes diversity within species, between species, and of ecosystems.

**Canadian Standards Association (CSA)** An accredited standards development organization established in 1919 and a non-profit voluntary association with focus on the development of a range of standards and product certification. The CSA sustainable forest management programme was developed for forest management operations in Canada.

**Certificate of conformity** A document issued under the rules of a certification system indicating the confidence that a duly identified product, process or service is in conformity with a specific standard or other normative document.

**Certification body** A unit that conducts certification of conformity.

**Certification** A process whereby an independent auditing body (third party) conducts an inspection and awards a certificate using independently developed standards and objectives.

**Certification of conformity** An action by a third party demonstrating the confidence that a duly identified product, process or service is in conformity with a specific standard or other normative document.

**Certifier** An organization that carries out the process of certification of assessment of conformity of another company to given standards.

**Certification programme** A system that has its own rules of procedure and management for carrying out certification of conformity.

**Chain of custody** Monitoring process of the production and distribution channel from forest to end product. It is the process of tracking and recording the possession and transfer of wood and fibre from a certified forest through the different stages of production—primary manufacturer, secondary manufacturer, wholesaler, and retailer—and to the end user.

**Civil society** A term used as an all-embracing term for voluntary associations that can be identified between the levels of the state, market, and individual citizens and their families. As such, civil society includes non-governmental organizations, non-profit associations, informal organizations addressing public interest issues, and self-help groups and associations.

**Community** Refers to all the people living in or adjacent to a forest in self-defined groups.

**Continual improvement** Year-on-year enhancement of overall environmental performance, not necessarily in all areas of activity, resulting from continuous efforts to improve in line with the environmental policy of the local forest management unit.

**Corrective action request (CAR)** A formal document raised by an assessor during assessment or surveillance, which details non-compliances identified and remedial measures required within a specified time frame.

**Criteria** Key elements that define principles. They are a means of judging whether or not a principle (of forest management) has been fulfilled.

**Customary rights** Result from a long series of habitual or customary actions constantly repeated, which have by such repetition and uninterrupted acquiescence acquired the force of a law within a geographical or sociological unit.

**Directory of registered organizations** Comprises formal documentation usually published annually by a certification body, which details those organizations currently holding certifications awarded by that body. It may also detail information provided on the related assessment schedules.

**Ecosystem** A community of all plants and animals and their physical environment and functioning.

**Endangered species** Species in danger of extinction throughout all or a significant portion of its range.

**Environment** Includes the forest surroundings and conditions in which a local forest management unit (LFMU) operates, including living systems (human and other). As certain environmental effects of the LFMU may reach all parts of the world, the environment in this context extends from within the LFMU to the global system.

**ENGO** An environmental non-governmental organization (NGO), such as World Wide Fund, Greenpeace, Conservation International or the Environmental Investigation Agency. The goals of environmental NGOs include, but not limited to, creating relationships with the government and other organizations; offering training and assistance in agricultural conservation to maximize the use of local resources; establishing environmental solutions; and managing projects implemented to address issues affecting a particular area. ENGOs are not run by federal or state governments but have funds issued to them by governments, private donors, corporations, and other institutions. To fully understand the social, economic, and environmental effects an organization can have on a region, it is important to note that the organization can act outside the formal processes that state governments and other government institutions must comply with.

**Environmental management system (EMS)** Indicates the organizational structure, responsibilities, practices, processes, and resources to implement environmental management.

**Environmental objectives** Broad goals arising from the environmental policy that an organization sets itself to pursue, which are quantified wherever possible.

**Environmental policy** A public statement of the intentions and principles of action of an organization regarding its environmental effects, giving rise to objectives and targets.

**Environmental targets** Detailed performance requirements, quantified wherever practicable, applicable to organizations or parts thereof and derived from the environmental objectives and that need to be met to achieve those objectives.

**Exotic species** An introduced species not native or endemic to the area in question.

**First-party verification** Where a company makes a statement about its own forest management practices based on its own standards and objectives.

**Forest integrity** Refers to the composition, dynamics, functions, and structural attributes of a natural forest.

**Forest management plan (FMP)** A base document adopted to describe the principles upon which forest management units will be managed.

**Forest management/manager** Refers to the people responsible for the operational management of the forest resource and the enterprise, as well as the management system and structure, and the planning and field operations. The forest manager is the person principally responsible for the activities of local forest management units.

**Forest Stewardship Council (FSC)** An international non-profit organization that offers forest certification internationally.

**Genetically modified organisms (GMOs)** Biological organisms induced by various means to develop genetic structural changes.

**Governance** Refers to the structure for overseeing and maintaining a certification standard. The governance structure of forest certification systems is generally made up of a diverse list of interested stakeholders.

**Governing body** Oversees the assessment and certification activities of the certification body and has equal and impartial representation of all interested parties.

**Guidelines** Overall directives that encompass and generally describe principles and criteria.

**High conservation value forests (HCVFs)** Possess one or more of the attributes defined by the Forest Stewardship Council's Principles and Criteria of Forest Stewardship.

**Indigenous lands and territories** Include the total environment of the land, air, water, sea, sea ice, flora and fauna, and other resources that indigenous people have traditionally owned or otherwise occupied or used.

**Inspection body (for certification)** Performs inspection services on behalf of a certification body.

**Interested parties** Please refer to stakeholders.

**International Organization for Standardization (ISO)** A network of the national standards institutes of 146 countries. ISO establishes generic standards applicable across all industry sectors.

**Lead assessor** Someone who is qualified and authorized to manage certification assessment.

**License (for certification)** Refers to the authorization of a certification body to grant a license.

**Local forest management unit (LFMU)** Forest area to which the certification applies.

**Long term** Refers to the timescale of the forest owner or manager as manifested by the objectives of the management plan, rate of harvesting, and commitment to maintain permanent forest cover. The length of time involved will vary according to the context and ecological conditions and will be a function of how long a given ecosystem takes to recover its natural structure and composition following harvesting or disturbance or to produce mature or primary conditions.

**Mark of conformity (for certification)** Protected mark, applied or issued, under the rules of a certification programme, indicating adequate confidence that the relevant product, process or service is in conformity with a specific standard or other normative document.

**Market-based instrument** An economic tool that affects costs and benefits of alternative actions open to forest managers and commercial and industrial stakeholders, with the effect of influencing behaviour in a way favourable to the forest environment.

**Montreal Process** An international working group established to identify and define criteria and indicators for the sustainable management of temperate and boreal forests.

**Mutual recognition** A process whereby an entity states that multiple certification systems meet its definition for sustainable forest management. The driving force behind mutual recognition is the existence of multiple certification systems in the market and confusion over whether or not they guarantee a certain level of sustainable forest management.

**National forest service** National government forestry organization responsible for regulation of activities in the forestry sector.

**Native species** Occurs naturally in the region and is endemic to the area.

**Natural cycles** Refer to nutrient and mineral cycling as a result of interactions between soil, water, plants, and animals in forest environments that affect the ecological productivity of a given site.

**Natural forest** An area where most of the principal characteristics and key elements of native ecosystems such as complexity, structure, and diversity are present, as defined by the Forest Stewardship Council-approved national and regional standards of forest management. A natural forest is any forest or closed woodland resulting from natural regeneration, consisting almost entirely of native species. It has developed with little to no human management of the forest stand.

**Non-industrial private landowner** Has less than 10,000 contiguous acres of forestland and is not affiliated with a forest product manufacturing facility. Note that this definition applies to the USA and may vary in other parts of the world.

**Non-timber forest products (NTFPs)** Refer to the goods and services produced from the forest covered by forest management units other than wood.

**Other forest types** Areas that do not fit the criteria for plantation or natural forests and are defined more specifically by the Forest Stewardship Council-approved national and regional standards of forest management.

**Peer review panel** An independent committee set up and monitored by the governing body to examine and comment upon assessment reports prior to final certification and registration of applicant organizations.

**Permanent forest estate (PFE)** Area of forest in a particular country designated and demarcated as permanent, usually divided into production and protection forests and subcategories of these two groups.

**Plantation forest** An area planted with trees of one or more species usually, but not exclusively, for wood production.

**Plantation** A forested area managed principally for wood production commonly consisting of a single species of the same age. A plantation lacks most of the principal characteristics and key elements of native ecosystems as defined by the Forest Stewardship Council-approved national and regional standards of forest stewardship. It results from the human activities of planting, sowing or intensive silvicultural treatments.

**Policy document** First-level documentation of the formal internal management system operated by a forest management unit (FMU). This document states the policy of forest management to which the FMU is committed. It includes an environmental policy. It can also include other policy commitments such as a quality policy and confidentiality policy.

**Precautionary principle** Action to prevent possible incidences of environmental or other damage, without waiting for scientific proof of the possible cause or effect of such damage.

**Principles** Key elements of a standard that define its scope.

**Procedures manual** Second-level documentation of the formal internal management system operated by a forest management unit (FMU). This document contains the operating procedures of the FMU, including the forest management plan.

**Product label** A mark affixed to the product providing information such as the content of certified raw materials in the product. The certification systems usually have labels distinguishing products that use 100% certified content and those with a certain percentage of content.

**Production forest** Area of natural and plantation forest within the permanent forest estate designated to be managed for productive purposes in perpetuity.

**Programme for the Endorsement of Forest Certification (PEFC)** An umbrella organization designed for mutual recognition of forest certification systems throughout the world.

**Quality forestry** Forestry that is environmentally sensitive, socially aware, and economically viable.

**Registered organization** A forest management unit that has satisfactorily undergone the process of assessment and certification and has been formally entered into the directory of registered organizations.

**Second-party verification** Where an affiliated body, such as a trade association, certifies its members using collectively developed standards and objectives.

**Silviculture** Art of producing and tending a forest by manipulating its establishment, composition, and growth to best fulfil the objectives of the owner. This may or may not include timber production.

**Smallholder** An individual or family that owns and relies on a small area of farm or forestland for subsistence and/or income.

**Stakeholders** Individuals and organizations who have a legitimate interest in the goods and services provided by a forest management unit (FMU) and in the environmental and social effects of an FMU's activities, products, and services. They include organizations exercising statutory environmental control over FMUs. Stakeholders also include local people, employees, investors and insurers, customers and consumers, environmental interest groups, and the general public.

**Standards** Documented agreements containing technical specifications or other precise criteria to be used consistently as rules, guidelines or definitions of characteristics to ensure that materials, products, processes, and services are fit for their purpose.

**Standardization** A process of developing, maintaining, and enhancing the standards by which the performance of an applicant for certification can be assessed.

**Stewardship** Involves managing forests with due regard for the long-term and external effects of activities, assuming responsibility for actions, accepting accountability, and adopting principles of good husbandry.

**Succession** Refers to progressive changes in species composition and forest community structure caused by natural processes (non-human) over time.

**Supplier** Party responsible for the product, process or service and able to ensure that quality assurance is exercised. The definition may apply to manufacturers, distributors, importers, assemblers, service organizations, and so on.

**Supply chain** Refers to the chain of ownership through which a product passes, for example, from the forest to the timber processor, manufacturer, importer, distributor, and retailer.

**Surveillance** Periodic re-examination and validation (see assessment).

**Sustainability** Refers to a state or process that can be maintained indefinitely. The principles of sustainability integrate three closely interlinked elements—environment, economy, and social system—into a system that can be maintained in a healthy state indefinitely.

**Sustainable Forestry Initiative** A forest certification programme designed to document sustainable forest management on industrial lands in Canada and the USA.

**System-based assessments** Such as those established under the International Organization for Standardization (ISO), inspect and certify a company's management systems, which allows a company to monitor its environmental management and impact.

**Tenure** Refers to socially defined agreements held by individuals or groups (either recognized by law or customary norms) on the rights of access and the rules for use of either a land area or associated resources such as individual trees, plant species, water or minerals.

**Use rights** Rights given for the use of forest resources that can be defined by local customs, mutual agreements, or prescribed by other entities holding access rights. These rights may restrict the use of particular resources to specific levels of consumption or particular harvesting techniques.

**Verification** Takes place through an audit, which can be external or internal. An external audit is carried out by an independent third party. It may be preceded by an internal audit by the organization itself (first party) to ensure that compliance with the requirements set can be successfully verified during the external audit.

**Wood procurement system** Refers to an auditable system that assesses the overall characteristics of raw wood and fibre from forest to end user.

# Index


**A**

Accreditation Services International (ASI), 48, 144

African Timber Organization (ATO), 16, 141

*Agenda 21*, 13, 14

American Forest and Paper Association (AF&PA), 8, 33, 50

American Tree Farm System (ATFM), 33, 118, 139

**B**

Ballarpur Industries Ltd (BILT), 75, 76

Bhopal India process, 22, 23, 223,

biodiversity conservation, 97, 116–118, 123, 126–127, 131, 135, 136, 155

- concerns in certification system for, 128–129
- forest certification on, 135–136
- threats to, 128
- ways to improve, 127–128

biological diversity (biodiversity), 5, 13, 19, 47, 59, 74, 129, 154, 219

- definition of, 128
- impact of certification on, 130–131
  - limitations in assessing, 131–132
  - measuring of, 130–131
- improvement in, 130–131
- and sustainable forest management, 126–127

BRE environmental assessment method (BREEAM), 198

*Brundtland report*, 1, 4

Brundtland, Gro Harlem, 4

buyers groups, 71, 72, 176

**C**

Canadian Standards Association, 10, 33, 36, 45, 55, 128, 140

carbon offsets, 150–156

carbon regulatory mechanism, 154–156

- scope and impact of, 154
- and SFM, 155–156
  - future of, 156
  - similarity between, 153–154

carbon sequestration, 137, 144, 145, 150–153, 155

Centro, Julio Cesar, 179

certified forest products, 27, 65–78, 101, 137, 163, 165, 166, 176, 179, 186, 194, 197, 199, 202

- assessment of new markets for, 66
- demand for, 66–67
- European markets for, 70–71
- feedback on, 69–70
- growth of marketplace for, 65, 73–74
- market demand for, 67–68

chain of custody (CoC), 28, 35, 48, 56, 60, 61, 66, 79, 87, 89, 105, 108, 150, 165, 185, 204, 210, 222, 227, 228, 229

- certification process of, 87–88
- defined, 88
- main components of, 88
- need for, 89–90
- and requirements of PEFC and FSC, comparison of, 92–93
- verification of, 93

clean development mechanism (CDM), 154, 155
clear-cutting, 10, 215
climate change, 4, 13, 145, 149–152, 157, 166, 198
- and forest certification, link between, 151–152
- mitigation of, 145, 149, 152
- and risk management, 152–153
- and SFM, link between, 150–151

CoC certification, 67, 81
- cost of, 108–109
- extent of, 67
- three things needed for, 80

CoC standards, 67, 68, 191
- developments in, 68

corrective action request (CAR), 85, 123, 130
- types of, 85

D

Dovetail Partners, 48, 51, 55

E

Earth Summit, 4, 6, 13, 128
- forest principles of, 4, 84
- outcome of, 13

eco-labelling, 50, 72, 139, 165, 207, 209
- misleading claims of, 139, 189,
- types of, 72

Economic and Social Council (ECOSOC), 6
environmental sustainability, 9
Eu-FLEGT process, 204
European business-to-business markets, 71–72
- characteristics of, 71–72

European consumer markets, 72–73

F

Fairtrade–Forest Stewardship Council (FSC), 144
FMU certification in India, 222
forest,
- ecosystems, 5, 10, 139
- ethics, 138, 139, 189
- quest for holistic certification of, 144
- management, certification of, 65, 81, 106
- resources, accreditation of, 153–154

forest certification,
- achievements of, 115–116
- analysis of costs of, 108–109
- and biological diversity, 219
- case study on, 84, 122–123
- challenges to, 175–184
- choice of programmes for, 63, 145–146
- and climate changes, 151–152
- components of, 27, 28
- consequences of high cost on, 109–110
- a conservation tool, 127–128
- convergence between, 141–143
- costs and benefits of, 111
- customer expectations from, 101–102
- design and surveillance, 86–87
- direct costs of, 105, 106

divergence in, 138
doubtful side of, 178–179
economics of, 105–114
effectiveness of, 120–121
effects of, 117–121, 135–136
analysis of, 118–119
on deforestations, 120
on forest management, 121
on plantation management, 135–136
elements of, 79, 81
emergence of, 31–34
in America and Europe, 33–34
enhancing the performance of, 192–194
and expectations of the,
business groups, 101
common people, 100–101
environmental groups, 99–100
forest companies, 100
investors, donors, and insurers, 103
manufacturers, 100
retailers, 99
stakeholders, 95, 99–101
factors limiting the spread of, 175–179
fire management by, 131
FSC approach for, 30–31
future of, 156, 185–202
challenges to, 185–186
threats to, 192
harmonization needs for, 191
implementation by forest owners, 217–218
impact on plantation management, 135–136
important ingredients in, 85–86
in India, 74–75, 164–165
strategy for, 165–166
improvement in fire management by, 131
indirect costs of, 105–107
inflexibility of standards in, 118–119
initiatives to combat illegal logging, 34–35
integrity and ethics needs in, 189
international schemes for, 45–54
issues and emerging trends in, 170–172
limited growth of, 179
main drivers of, 194–195
main impacts of, 116–117
management goals for, 160
market advantage of, 161
market driven approach to, 28, 42–43
national initiatives for, 55–60, 74–75, 165–170
objectives of, 29–30
opportunities and limitations of, 186–187
phased approach to, 222–223
pitfalls and solutions to, 93–94
positive impact on plantation of, 122
prime objective of, 125
process of, 79–94
three things needed in, 80
standards of, 59, 62
proliferation of, 137–138
drawbacks of, 137
regulatory mechanism for, 44–45
salient features of, 59
slow progress of, 175–176
and social responsibility, 43–44

and stakeholder consultation, 51, 81, 83, 86, 93, 94, 106, 142
stakeholders interests in, 97–98
and schemes for, 41–63, 143–144
status and extent of, 35–40
types of, 80

forest certification schemes, 45, 52, 59, 63, 110, 131, 138, 140, 210, 214, 219
comparison of, 59
five possible avenues for, 141
harmonization for, 141
war for market share between, 140

forest certification standards, 42, 140, 143, 165, 170, 196
and the national laws, 140
conflicts between, 140–141

forest management unit (FMU), 16, 23, 28, 56, 74, 81, 97, 98, 106, 119, 123, 132, 182, 215, 217, 222, 227, 229, 231

Forest Stewardship Council (FSC), 10, 27, 36, 41, 45, 46, 50, 60, 67, 81, 84, 99, 108, 111, 112, 116, 123, 128, 130–131, 138, 142, 144, 146, 151, 168, 185, 210, 214, 219
activities of, 46–47
background of, 46
operational structure of, 47
and PEFC system, 138
discord between , 138
principles and criteria of, 47–48
relevant principles of, 130
trend of changes in, 48–49
work on pesticide use in Germany, 131

forest sustainability, 4, 139, 192–194
important landmark events in, 4–6

FSC–certified timber, 44, 71, 195

FSC–CoC certificates, 195
for paper industries, 195

G

genetically modified organisms, 10, 142, 145

global broadband and innovations programme, 44

global carbon cycles, 5

Global Forest and Trade Network (GFTN), 78

green building initiatives (GBIs), 34, 171, 197–200
examples of, 198

green globes, 198, 199

Green India Mission, 166

green premium wood, 25, 65, 178

green public procurement, 44, 194, 195, 197
efficacy of, 197

greenhouse gas (GHG), 149

group certification, 48, 110, 117, 144, 182, 186, 210, 223, 228, 229, 230

H

Helsinki process, 14

I

illegal logging, 7, 34, 78, 194
initiative to combat, 34–35

India,
accreditation and certifying bodies of, 224
certification of forests in, 160
community forest programmes in, 74, 76
draft national criteria and indicators for SFM, 125, 127

FSC certifications in, 76–78
Indian Forest Certification Council (IFCC), 162, 163, 220, 221, 222, 224, 228
constitution of, 228–229
objectives of, 229
main recommendations of, 162
task of, 163–164
Indian Institute of Forest Management (IIFM), 22, 170, 223
Indian working plan code, 172
industrial/irresponsible logging, 128
Intergovernmental Forum on Forests (IFF), 6
Intergovernmental Panel on Forests (IPF), 6
International Certification Standards, 196, 223
International Organization for Standardization (ISO), 34, 45, 50, 144, 206, 209
labelling standards of, 209
International Tropical Timber Agreement (ITTA), 17
International Tropical Timber Organization, 7, 16, 17, 19, 74, 129, 141, 216
ISO 14001, 34, 56, 61, 62, 108, 121

**K**

Kyoto protocol, 149–157
objective of, 154

**L**

Lacey Act, 44, 206
Leadership in Energy and Environmental Design (LEED), 52, 198, 199
legal timber, 35, 201
London Olympics 2012, 44, 195

**M**

Malaysian Timber Certification Council (MTCC), 56–58, 224
certifications of, 56
impact on the six core areas of, 58–59
institutional arrangement of, 57–58
Maya Biosphere Reserve (MBR), 131, 159
Menominee tribe, 6
Montreal process, 5, 14
criteria and indicators of, 5–6

**N**

National Certification Standards, 176, 191, 196, 223, 225
National Forest Certification Committee (NFCC), 162, 220
report of, 162
National Wildlife Federation, 133

**O**

Omnibus certification scheme, 143–144

**P**

pan-European,
forest process, 10, 14,
operational level guidelines, 17
process, 53
Pan-European Forest Certification Council (PEFCC), 54, 140
private non-industrial forests, 177
Programme for the Endorsement of Forest Certification (PEFC), 10, 17, 33, 36, 38, 42, 45, 51–57, 67–69, 138, 140, 142, 143, 145, 191, 195, 196, 198
-CoC certificates, 67, 68
criteria for, 53–54

and FSC, convergence between, 143
and process of, 54–55
regional certification approach of, 55

**Q**

quality forestry, 26, 27, 99, 178, 217

**R**

rainforest alliance, 32, 70, 135
Redd+, 151
aim of, 151
regulated forest, 3
Rio declaration, 13
riparian zones, 31, 122

**S**

Seymour, 189
social sustainability, 9
Softwood Lumber Agreement, 138
stakeholders' consultation, 51, 81, 83, 86, 93, 94, 106, 142, 221–222
sustainability,
characteristics of, 2
defined, 1
in forest management, 2
in Indian forestry, 7–8, 20–22
historical evidence of, 20–21
modern efforts for, 20–22
origin of, 2–3
sustainable forest management, 1, 4–11, 19, 22, 25, 26, 28, 32, 33, 40, 42, 50, 74, 78, 80, 95, 98, 110, 118, 126, 134, 137, 139, 149, 159, 163, 175, 182–184, 186, 188, 192, 197, 200, 223
accreditation under, 153
basic requirement for, 17–18
and carbon credit projects, 153–154
certification in India for, 170–172
considerations for, 172–173
and climate change, link between, 150–151
criteria and indicators for, 5–6, 17, 141
debatable facts about, 19–20
defined, 7
earlier attempts to, 8
effects of biodiversity on 134
essential aspects of, 8–9
genesis of, 12–13
goal of, 11
holistic view of, 189
Indian initiative for, 22–23
intergovernmental initiatives for, 14–15
objective of, 154
policies for, 189–191
regional standards for, 10
rise of, 4
Sustainable Forestry Initiative (SFI), 33, 41, 50–51, 111, 118, 128, 138, 139, 142, 185, 219
changes in, 51
impact of, 52
objectives of, 50
strategy at the national level, 203
sustainable plantation needs, 12

**T**

Tarapoto Proposal, 14
The Programme for the Endorsement of Forest Certification (PEFC), 191

**U**

United Nations Conference on the Human Environment, 13

United Nations Forum on Forests (UNFF), 6
United States Green Building Council, 52, 189
US Green Building Council (USGBC), 52, 139, 199, 208
US National Association of State Foresters, 139

**V**

voluntary partnership agreements, 204, 205

**W**

water monitoring, 122
Wolf river treaty, 7
World Commission on Environment and Development, 1, 4, 13
World Conservation Strategy, 1
World Forestry Congress, 187
World Wide Fund (WWF), 32, 46, 52, 71, 73, 74, 78, 224
  working group of, 224

# About the Authors

**Dr H S Gupta** is currently Professor in Technical Forestry at the Indian Institute of Forest Management, Bhopal, with a current research focus on collaborative working in forestry projects, plantation forestry, management of wildlife, and climate change in forestry. He is an Indian Forest Service officer of the Jharkhand cadre (1986 batch). Dr Gupta has done his PhD from BIT Mesra, MTech from IIT Delhi, and BSc from Allahabad University in Agricultural Engineering. He has worked extensively in different districts of Bihar and Jharkhand, carrying out various models of plantations/afforestation, forest and wildlife management, and has analysed the different dimensions of their efficacy and efficiency. His current interest and research experience are in climate change related issues to forestry, tribal and rural development, conflict resolution, policy, and laws in natural resource management, sustainability of natural resources and certification aspects of forestry.

**Dr M Yadav** is currently Associate Professor at the Indian Institute of Forest Management, Bhopal. He has a PhD in international marketing, MBA in marketing management, and graduated in Dairy Technology. He has 25 years of experience—nine years of industry experience and 15 years of academic experience. While in academics, he taught and undertook research in areas such as forestry products trade, WTO; trade and environment-related issues; sustainable forest management and forest certification; corporate environmental responsibility; and green consumerism. He has published a number of articles in peer-reviewed journals and has authored a monograph titled *Forest Certification: a tool for sustainable forest management*. He has been conducting training programmes on forest certification for various stakeholders, including Indian Forest Service officers.

**Dr D K Sharma** is DIG (Forests), Ministry of Environment and Forests, Government of India. He belongs to the 1988 batch of Indian Forest Service of Tripura, part of the Manipur–Tripura Cadre. He has served as Chief Conservator of Forests and Director, NTFP

Centre of Excellence. He completed his masters in Chemistry and his postgraduation in Public Policy and Management from IIM, Bangalore. His doctoral work is on forest fire management and policy issues and was awarded PhD by the FRI University. Dr Sharma has written many articles and books on protected areas governance, environmental issues, NTFP management, forest fire issues, and improvement in the forestry sector.

**Mr A M Singh** is an IFS officer of 1986 batch from Assam-Meghalaya cadre, presently serving as Inspector General of Forests in the MoEF dealing with welfare of elephants (Elephant Conservation and Management). Mr Singh had dealtForest Certification, Timber Trade; NTFP Management, Forest Cover Assessment and Utilization of forest Products as DIG (SU) in the Ministry of Environment & Forests as his last assignment. He had served more than 12 years in the state in different capacities as DFO in Territorial, Social Forestry and Working Plan divisions. He had also served as Associate Professor/ Professor in IGNFA for more than seven years from 1999 to 2007. He is havingexpertise in Sustainable Forest Management, Gender Issues in Forestry, Joint Forest Management, NTFP Management and Forest Certificationand presented many papers on these issues of forestry at National and International Conferences. IUFRO, World Forestry Congress, Commonwealth Forestry Conferences are few of them. He has developed modules on Wetland Management, Gender Issues in Forestry and Forests and People for forest officers. Last year, he wasnominated as one of the members in the expert panel of International Tropical Timber Organization (ITTO) for evaluation of forestry projects and representing Asian Producer Countries.